Electronic Circuits

The Definitive guide to Circuit Boards Testing Circuits and Electricity Principles 2nd Edition

Wayne Charles

Copyright © 2016 Wayne Charles

All rights reserved.

ISBN: **9781973307617**

☐ Copyright 2016 by Wayne Charles - All rights reserved.

This document is geared towards providing exact and reliable information in regards to the topic and issue covered. The publication is sold with the idea that the publisher is not required to render accounting, officially permitted, or otherwise, qualified services. If advice is necessary, legal or professional, a practiced individual in the profession should be ordered. - From a Declaration of Principles which was accepted and approved equally by a Committee of the American Bar Association and a Committee of Publishers and Associations. In no way is it legal to reproduce, duplicate, or transmit any part of this document in either electronic means or in printed format. Recording of this publication is strictly prohibited and any storage of this document is not allowed unless with written permission from the publisher. All rights reserved. The information provided herein is stated to be truthful and consistent, in that any liability, in terms of inattention or otherwise, by any usage or abuse of any policies, processes, or directions contained within is the solitary and utter responsibility of the recipient reader. Under no circumstances will any legal responsibility or blame be held against the publisher for any reparation, damages, or monetary loss due to the information herein, either directly or indirectly. Respective authors own all copyrights not held by the publisher. The information herein is offered for informational purposes solely, and is universal as so. The presentation of the information is without contract or any type of guarantee assurance. The trademarks that are used are without any consent, and the publication of the trademark is without permission or backing by the trademark owner. All trademarks and brands within this book are for clarifying purposes only and are the owned by the owners themselves, not affiliated with this document.

CONTENTS

	Contents	i
1	Introduction	ii
2	Chapter 1: Common Units Used	Pg 1
3	Chapter 2: Types of Circuits	Pg 4
4	Chapter 3: Practice Makes Perfect	Pg 10
5	Chapter 4: Combined Circuits and Measurements	Pg 13
6	Chapter 5: Types of Current	Pg 17
7	Chapter 6: Introduction to Circuit boards – The Breadboard	Pg 19
8	Chapter 7: Stripboard	Pg 23
9	Chapter 8: Printed Circuit Board	Pg 29
10	Chapter 9: Testing the Circuits	Pg 35
11	Chapter 10: Finding a Fault in a Circuit Board	Pg 39
12	Chapter 11: Mistakes to Avoid as an Electronics Beginner	Pg 42
13	Conclusion	Pg 44
14	Free Bonus	Pg 45

Wayne Charles

INTRODUCTION

Electrical circuits have become an integral part of our everyday lives, and nearly everything we touch these days has some form of electrical circuit in it. From the watch on your wrist or the smartphone in your hand to some of the toothbrushes that we use to brush our teeth, circuits are virtually impossible to avoid.

However, not very many people know or even understand what a circuit is and how it works. In this book, you shall not only find out what a circuit is and how it works, you will also learn how to test the circuits, and the basic principles of electricity.

People have known about electricity since the time of the Ancient Greeks. In fact, scientists seem to have found proof that the Ancient Romans had invented a primitive battery made from clay pots lined with copper. The scientists believe that the clay 'batteries' must have been used to produce light at Roman sites.

There are those who do not believe that the Ancient Romans could have been that developed, however, sites near Baghdad have shown that even the Persians were using a device similar to the Roman one. Perhaps the Persians and the Romans traded ideas when they realized how beneficial such a development could be.

However, one thing that most historians will agree on is the fact that the Italian physicist Alessandro Volta was the first man to construct a working circuit that actually carried a steady electrical current. He achieved this feat in 1800, and ever since then, human beings have been experimenting with circuits, trying to figure out new ways to make them better, smaller, and more reliable.

In this book, you shall be introduced to the world of electricity and electrical circuits, and learn why they are so important. To begin with though, you should start right at the beginning, with the principles of electricity.

CHAPTER 1: COMMON UNITS USED

The principles of electricity are as simple as they are elegant. However, before you begin to delve into these principles, it is important to understand what electricity is in the first place. The simplest way to define electricity would be to say that electricity is the movement of an electric current from one place to another. The movement of free electrons from one place to the other causes this electric current, therefore, the more free electrons a material has the better it will be at conducting electricity.

There are various parameters that need to be considered when you discussing the principles of electricity, the most important of which are Coulombs, Volts, Amperes, Ohms and Watts.

1. Coulomb

The coulomb (C) is the internationally accepted (SI) unit of electrical charge. The Coulomb is named after the famous French physicist Charles-Augustin de Coulombwho developed a law by the same name to explain the electrostatic forces of attraction and repulsion. Its symbol is C, and it is used to measure the amount of charge carried by a constant current along a conductor in one second.

It can be described mathematically as: $1 C = 1 A \times 1 S$

1 C is roughly 6.242×10^{18} or 1.036×10^{-5} mol Protons, while -1 C is just about 6.242×10^{18} electrons.

2. Volts

Named after the famous Italian Physicist Alessandro Volta, the Volt (V) is a unit of measurement used to describe the electromotive force (emf) required to drive a current of one ampere through a conductor with a resistance of one ohm.

To simplify this explanation, imagine you have a hydraulic circuit, with a faucet, a pump, a pipe and a trough. The pump helps to draw water into the pipe when the faucet is turned on, and the water flows through the pipe, out of the faucet and back into the trough where the whole cycle starts again.

The volt in this case would be the amount of pressure that the circuit would need to get the water through it.

The volt can mathematically be represented in a number of ways. Expressing it in SI units (the seven internationally accepted units of measurement) can be done as follows:

V = Potential Energy/Charge = N x m/coulomb =
Kg x m x m/ s2 x A x s = Kg x m2/ A x s3

3. Ampere

An ampere (A) or an amp is the SI unit for electric current. It is equivalent to 1 coulomb, and is used to describe the flow rate of an electrical charge. The ampere is named after André-Marie Ampère, a French physicist and mathematician that was central to the development of classical electromagnetism during the early part of the 19th century.

It is important to remember that the amp should never be confused with the coulomb. This is because the coulomb is a unit of charge, where as the amp is a unit of current, or the amount of charge moving through the circuit at any given time.

It should come as no surprise then to realize that should there be an increase in the number of charged particles passing through a specific point, then there should also be a rise in the amperes of current.

Amperes can be mathematically expressed as follows:
1A = 1 C/s

4. Ohms

The ohm (Ω) is a unit of measurement that is used to describe electrical resistance through any conductive material. It was named after the German physicist and mathematician, Georg Simon Ohm, who was a one of the pioneers in the study of electricity.

He developed a law by the same name (Ohm's Law) to describe the relationship between a conductor, the potential difference (voltage) applied to it, and the consequent electric current. The law describes the ohm as an electrical resistance in a circuit transmitting a current of one ampere when subjected to a potential difference of one volt.

Basically, it states that the ohm is the resistance to the passage of electricity through a conductor should the said conductor have an electric current of 1 ampere pass through it, and a potential difference of 1 volt.

The ohm can be mathematically represented as: $\Omega = V/A$ (This is also how you would express Ohm's law)

In most cases, the resistance of the conductor in ohm's law remains constant even when taken through a range of voltages and temperatures. Such conductors are called linear resistors, and must not be confused with thermistors, which have a variable resistance depending on different factors, most critical of which is temperature.

5. Watts

The watt (W) is the SI unit used to measure power, and was named after the Scottish engineer James Watt. Electrical power is a gauge of how long it will take an electrical circuit to transfer electrical energy, and it is measured in joules per second.

The electric power generated by a circuit can be defined in three different ways:

1. The power lost in a circuit is inversely proportional to the squared value of the voltage travelling through it.

$P = V^2/R$

2. The power lost in a circuit is directly proportional to the squared value of the current going through it.

$P = I^2 \times R$

3. The power in a circuit is equal to the voltage multiplied by the current travelling through it.

$P = V \times I$

Where: P = Power
V = Voltage
I = Current
R = Resistance

6. Ampacity

Ampacity can be simply defined as the amount of current that a conductor can handle before its temperature begins making it sustain progressive or immediate damage.

Though it is not regularly considered when creating basic circuits, in larger installations the ampacity of a conductor should be considered as it can drastically affect the materials used to create the circuits.

The ampacity of a conductor is determined by:
1. The resistance of the conductor
2. The ability to lose heat
3. Ambient temperature
4. The insulation of the conductor

CHAPTER 2: TYPES OF CIRCUITS

In order to fully understand what makes up an electronic circuit, it is essential to define a circuit. An electronic circuit brings together a range of electronic components. When these components flow together, then they allow for an electric current. Typically, they are made up of two terminals are more for their to be a circuit diagram design, which will form a loop. When these components are soldered onto a circuit board, then the result is an electrical system.

Different elements of an electrical circuit can be connected in countless different ways. For this reason, before you learn about the different circuit boards there are and how they work, it is important to know what kinds of circuits there are, as many circuit boards contain one or more of these circuits. There are two main types of circuits, Parallel circuits and series circuits.

1. Series Circuits

Series circuits are so called because the components in the circuit are connected in a single line or series so that the same current flows through them uninterrupted. This means that all the elements of a series circuit carry the same current. An example of a series circuit is shown below:

One of the major advantages (or disadvantages depending on the application) of these circuits is, because all the components are connected in a straight line, breaking the circuit at any given point will cause the entire circuit to open or break.

The very nature of series circuits allows the calculations that have to be carried out for these circuits to be rather simplistic. For instance, should you want to calculate the resistance for 3 resistors that are arranged in series, the equation would be:

R¬¬¬Total = R1 + R2 + R3

This means that if you have three 4 Ω resistors then the total resistance is going to be:
RTotal = 4 + 4 + 4 = 12 Ω

Calculating the potential difference across the circuit is equally simple. For instance, should there be two 1.5 V batteries connected to the circuit in series, the total potential difference will be 3V, i.e.
Vtotal = V1 + V2

Series circuits are not commonly used due to the fact that should any component in the circuit fail, then the whole circuit fails until the issue with the component can be resolved. However, they are used in certain appliances such as torches, where the voltage supplied needs to be relatively high.

12V car batteries are also an example of components that are connected in series, as they are often two 6V batteries that are connected together. In trucks, two of these batteries are often connected in series to supply the 24V that the truck needs to run.

2. Parallel circuits

Parallel circuits are those where the potential difference (Voltage) across all the components connected to the circuit is the same. This fact also means that all the devices in the circuit will have the same polarities. However, this means that the current that flows through all the components of the circuit will be shared, meaning that the total current in the circuit will equal the sum of the currents through each component of the circuit. An example of a parallel circuit is shown below:

Calculating the resistance for a parallel circuit is slightly more complicated. To calculate the total resistance for the whole circuit, you have to add the reciprocals of all the resistances in the circuit, then take the reciprocal of the sum. Mathematically, that would be represented as:
1/Rtotal = 1/R1 + 1/R2 + 1/R3

If only two resistors are connected in parallel, then the equation becomes:
R¬total = R1R2/R1+R2

In a parallel circuit, the total resistance of the circuit always has a lesser value than the one of the smallest resistor. For example, if you have 4 10 Ω resistors, then your total resistance will be:
1/Rtotal = 1/10 + 1/10 + 1/10 + 1/10
1/Rtotal = ¼
Rtotal = 4

Parallel circuits have very many uses, and are appear often in everyday life. For instance, almost everything in your house is connected in parallel to the mains supply, as are most premises that are connected to the power grid. Parallel circuits are also used in Christmas lights and Chandeliers to ensure that even if one light goes off the others will still function.

3.   Series Parallel Circuit

It is possible to find a circuit that combines both the series and the parallel circuit, and this is known as the series parallel circuit. With this circuit, the electric current will travel through both of the circuits together.

When looking at all these circuits, it will become clear that there are three main elements that are involved. The first is the voltage source which enables the current to flow like it would in a battery. The second element is the load, which is where most of the work happens within the circuit. There are various things that can represent the load, such as a light bulb in a simple circuit. When the circuit becomes more complex, then there are other components that can be the load such as capacitors, resistors and so on. Finally, there is the third element, which is the conductive path. This enables the current to flow down a specific route. The route will have its starting point at the voltage source, then the journey through the entire load, and finishing up at the voltage source again. This means that the path that the electricity follows will be from the negative side of the source of voltage, all the way to the positive side of the source of voltage.

In circuits, there are distinct differences between an open circuit and a closed circuit. A closed circuit is one which forms a complete loop that enables the current to flow from one end to the other. This means that if there is any disconnection within that loop, then it is not possible for the current to flow. That is when you will find the next type of circuit which is known as an open circuit.

It may seem strange that an open circuit is so called, because the main function of a circuit is for all the components to form a path that is complete. However, if there is something that has erred in the circuit, then it can be described as open. This means that one needs to look at all the connections, or evaluate whether there is any damage to the circuit.

If you have a short circuit, then the entire circuit does not have a viable load. When this happens, it could be a significant indication of danger, so you should not ignore a short circuit. With a short circuit, the flow of the current can be much higher than normal. This means that it is possible for explosions of batteries to occur, or in the worst case, for a fire to begin. The electronic components that you are working with are also likely to get damaged.

What then is a short circuit?

Simply put, a short circuit is a low resistance connection that can occur between two conductors that supply electricity current to any circuit. This results into an excess of current flow from the power source through the short. This can cause a great damage to the power source. When this happens and there is a fuse in the supply circuit, it will blow up in order to open the circuit to stop the excessive current flow.

A short current can either be an alternate current (AC) or a direct current (DC). If for instance the short happens on the battery, it will be quickly discharged and at the same time heat up so much because of the high flow of the current.

A short circuit current on the other hand is the heavy current that flows to the circuit once the short circuit occurs.

Short circuit is however not an overload, as many people think. They both cause the same amount of damage to the system but they are very different. During a short circuit, the voltage at the fault point goes down to zero and very high current start flowing through the network to the fault point. An overload on the other hand means that the load of current is more than the values designed for a particular system. This means that the voltage is low but at least not zero.

Causes of a short circuit
There are both internal and external causes of a short circuit:

i) Internal causes: this means that a short circuit can occur in case of a crash in the equipment or even the transmission lines as a result of a weakening of the insulation in a transformer or even a generator. If the transformer or generator starts ageing, its insulation can start ageing too and this can result into a short circuit. The insulation can be of an insufficient design or the insulation may not be properly made.

ii) External causes: lightning rushes can also cause an insulation failure, causing a short circuit. Overloading of equipments is also a great cause of short circuits as it causes an excessive heating on the circuit, causing a great damage in the end. There are some faults that are caused by people too, which can result to a short circuit.

Effects of a short circuit
- The heavy current which results after a short circuit causes an overheat in the system, which can consequently cause fire and explosions. These will in turn cause great damage to the system.
- Short circuits also cause a low voltage in the system, which can affect the service which has been rendered by the power system. If the voltage stays at low for a few more seconds, great damage can be caused on the generators and other appliances that are using the electric power.

It is therefore very important to disconnect the faulty connection as fast as possible in order to prevent any damage for happening.

Electronic Circuit Symbols

When learning about electronic circuits, you will notice that there are numerous diagrams that you will need to refer to. It becomes impossible to understand these diagrams if you have no idea what the symbols mean. Therefore, this section shall illustrate these symbols to, particularly the most basic ones.

1. Wires

This is the most basic of the symbols and will appear in all circuit diagrams. It is through the wire that power is able to flow in the entire circuit. The wire will also connect different components and there are three types that you will encounter.

The single wire appears as a line that with two terminals. The terminals depict the starting of the wire and the ending of it as well. This is how the current is able to pass from one component to the next.

There are circuits where two or more wires are connected to each other. When this occurs, it is referred to as wires jointed. Where the wires join there is a small circle which is also known as the blob.

It is possible for two wires to appear to cross over one another, and this is what you will find in some circuits that are more complex. When this happens, they are not actually connected and what appears in the illustration is bridging.

2. Resistor

When you have electricity flowing through the wires, it is possible that you may want to interrupt it or resist it in some way at a point along the current. That is when you will use the resistor, with the symbol as illustrated. This can also be revealed as a box that appears within the wire. It is an element that is passive in nature, and it will oppose the flow of the current within a circuit. It typically has two terminals within it, which work together to bring to a halt the energy that is flowing, using heat as its main form. If there is an overflow of an electric current that passes through the resistor, it is likely to get damaged and change color.

3. Light Bulb

There are times that a current will pass through your electrical circuit, and the result will be that there is a bulb which glows. When you know that you have a light bulb in place, you can indicate it with the symbol as illustrated above. When the electrical current is not flowing, then the bulb will not glow.

4. Power Supplies

For an electronic device to function well it needs to have some sort of power supply so that energy passes through it. The power supply has as a primary function, which is the conversion of energy from one form to the next. Since there are different amounts of energy that can flow within a circuit, there are difference power supplies as well.

The cell circuit symbol as illustrated above allows electrical energy to be supplied from a large terminal that has a positive sign. The battery circuit shows that there are two or more cells that are involved. It works in the same way as the cell circuit.

5. Switch

When you need to connect electrical circuits, then you need a switch. This is a component which will facilitate this connection of circuits when it is closed, and when it is open, then that connection on the electrical circuit will be broken. There are a range of switches that you will find on electrical circuits, and the more complex your electrical circuit, the more switches that you will find.

6. Capacitors

This is also known as a condenser. From the illustration, you will note that it has two terminals which are passive in nature. These have the capacity to store energy that is then released as electricity. When supplying power, these can often be referred to as rechargeable batteries for the supply of power. Within capacitors, AC signals are easily filtered whereas DC signals are blocked.

There are three main types of capacitor. The first has been explained above. The second is the polarized capacitor which is able to store electrical energy as well. However, it is only able to this coming from one main direction. The variable capacitor on the other hand is adjustable, and with a small knob, they can be controlled.

7. Diode

This is an electrical component that has two terminals. One is known as anode, the other cathode. The diode enables the electron current flows, coming from the cathode to the anode. However, it is blocked when it is trying to flow from the other direction. This basically means that from one direction the level of resistance is low, but from the other direction, the resistance level is high. There are several different types of diodes that you will encounter, which also affect the way that one can tell an electronic current is flowing (normally with light), when the voltage has broken down, and in the instance that operations are at elevated speed.

CHAPTER 3: PRACTICE MAKES PERFECT

Before moving forward and finding out about more complex circuits, it is essential that you master the basics. This chapter will guide you through how to start creating your own circuits, beginning with the most simple that you can achieve. By taking the time to practice, you will find that it is much easier than you have imagined, and you can then develop on a solid foundation.

Light a Bulb
This is quite possible the easiest circuit that you will encounter, and you need minimal materials to get it working. Put together the following: -
- Copper Wires
- Battery
- 6-volt incandescent light bulb

Connect the wires to the battery, one to the positive end and the other to the negative end. Then connect these wires to the bulb, by wrapping them around the screw on section of the bulb. The bulb should immediately light up. Really, it is that simple. Here is a detailed guide:

A simple electric circuit needs a power source, which is a battery in this case, wires and a resistor, which will be a light bulb. For it to work, electrons will be flowing from the battery, going through the wires and then into the light bulb. When the electrons are enough, the light bulb will light up. The making of a simple circuit will not take so much time, and it is very easy.

    a)   Start by gathering all the materials you will need- like mentioned above, you will need a battery or a packet of batteries, two insulated wires, a light bulb and a light bulb holder. The amount of power the battery you will choose emits should guide you on the kind of bulb you will buy.

    b)   Cut the ends of the isolated wires- this is in order to expose the wires, for them to work well for the experiment. Remove some bit of the insulation from the ends of the wires in order to expose parts of the wires on both ends.

c) Fit the batteries into the battery pack, ensuring that the positive ends and the negative ends are in alignment. Ensure that all the batteries are placed close together and hold them in position using a battery pack or anything else you might have.

d) Now attach your wires to the battery pack- these wires will be used to pass electric current to the light bulb from the batteries. To be safe, you can use an electric tape to fix the wires securely on each battery. You cause a battery snap as well. Be careful because it is very easy to get an electric shock if you touch a naked wire that has already been connected to a battery.

e) Once done with all the batteries, fix the other end of the wire on the metal part of your light bulb. Loosen each screw on the light bulb and fix one end of the wire around each of the screws. Fasten the screws in order to hold the wires in position.

f)Test the circuit- place the light bulb on your light bulb holder and ensure that it is in position. Your bulb should light at this instant if your connection was done well. If it does not light up, you might have to go back to the beginning to check whether the connection was done properly.

Another simple electric circuit

This is a simple exercise that should help every beginner out there to learn how to generate current, which they can then use in different areas.

Materials needed:
- A battery or a coin cell- you do not have to buy this as they are present in most of the toys children play with, in watches, calculators and other small electric devices.
- Scissors-
- A cardboard-
- Zip locks
- A two-sided tape-
- Resistors- this is optional
- LED- found in most toys and electronic devices too
- An aluminum foil
- Glues or adhesives
- Decoration materials- for decorating and labeling your project

Give your circuit a shape

Cut the cardboard into the shape you want your circuit to take. Wrap it up with a color paper, choosing probably a dark color for the background.

Make c-shaped sets using the two sides tape that have been cut into small rectangle-shape pieces. Make two sets of these, facing the opposite directions and ensure that you stick them at a distance from each other. If you want to create a switch, keep one of the sets longer than the other.

Place the aluminum strips onto the c-shaped tape strips. Peel off the plastic later from the tape and the stick the conductive strips onto the tape strips. For the switch strip, cut the foil a little longer so that it can wrap around the switch strip. The foil should be intact onto the c-shape and it should not break at all. In case of a breakage, the current will not flow through it; therefore, the project may not be a success.

Create a switch

You can use a coin or a button to create a battery holder. To do this, cut a circular ring on a thin cardboard, ensuring that its inner circumference is more than the circumference of the battery that will fit into it so that the battery will fit perfectly inside the battery holder. Stick this circular ring onto the c-shape set you created for the switch.

Now place the coin or the button inside the holder.

Finish the circuit

- Place your LED on the opposite side of the switch and keep them in place using paper strips or a transparent tape. Ensure that the LED is paced in the right direction. To get it right, you can test it directly on the battery first.

- Now test your circuit. You do this by pressing the switch over the battery.

- To test if the current can be conducted from one object to another, use conductive materials like materials made from nickel, brass or steel. You can use copper wire too, or any object made of iron. Try with non-conducive materials too to test whether the current can flow through them.

- Finally, you can decorate your project to make it look more alive.

CHAPTER 4: COMBINED CIRCUITS AND MEASUREMENTS

An electrical circuit may not always be either series or parallel. A lot of the time, both circuits are combined to form a compound circuit. Shown below is an example of a combined circuit:

AWhen confronted with some of these circuits, the best thing would be to analyze them and simplify, especially if you have to work out the resistance, voltage, or current for any given component of the circuit.

For example, if you have a circuit with 4 resistors, three connected in parallel (10 Ω, 20 Ω and 30 Ω respectively) and one in series (10 Ω) and you have been asked to calculate the total resistance in the circuit, the first thing you should do is calculate the total resistance of the resistors in parallel. Once you have done this, you can then add up the resistors that are in series and find the total resistance, as illustrated below:

1. Add up the resistors in parallel
1/Rtotal = (1/10 + 1/20 + 1/30)
1/Rtotal = (6 + 3 + 2)/ 60
1/Rtotal = 60/11 = 5.45 Ω

2. Add the replacement resistor (5.45 Ω) to the final equation (Rtotal = R1 + R2)
Rtotal = 5.45 + 10 = 15.45

Connecting Instruments to your Circuit

When you finally start experimenting with circuits, you will realize that many of the tests you will be carrying out will be done live. To ensure that you take the most accurate readings possible, it is important to pay attention to this section before you move on.

There are certain things that are worth remembering when you are about to start taking measurements. One is that 'Current flows through a conductor'. A conductor is any material that allows electricity to pass through it easily. Most metals are conductors, as is water and any other compound that easily shares electrons.

On the other hand, insulators do not share electrons very easily and therefore are considered bad conductors of electricity. Good insulators of electricity include materials such as glass, plastic, rubber and wood.

There are a few things you need to consider before you connect a meter to your circuit including, but not limited to the following:
- Do I have the right device for the job?
- Is it damaged in any way?
- Do I know how to calibrate it, or how it works?
- What am I trying to test, and therefore, how should I connect it to the circuit?

Well, some of these questions you are going to have to answer for yourself. However, in regards to the last question, this section should help you a great deal.

Measuring current with an ammeter

Measuring current is very simple if you have an ammeter. The fact that current needs to flow through a load means that for an ammeter to display a reading, it needs the current to be directed through it. For this reason, an ammeter must be connected in series as shown in the diagram below.

If you are measuring small currents this should not be an issue. However, for larger currents, especially if you do not know what type of values you will be expecting, you need to set your meter on the largest current range. If you get a smaller reading that expected, then you can slowly turn down the meter to get a more accurate reading.

Remember, to ensure that you get an accurate reading your ammeter needs to have a very small internal resistance otherwise both the overall resistance, and therefore the current flowing through the meter will be affected.

Measuring voltage with a voltmeter

As you know, the voltage or potential difference of a circuit is the amount of electromotive force (emf) required to drive a current round a circuit. Therefore, a potential difference will exist across a load to ensure that the electrons go through the load. To measure this potential difference across a load, the voltmeter is connected parallel to the said load as shown below

Unlike the ammeter that has a very low internal resistance, the voltmeter is the opposite, and instead has a very high internal resistance. The voltmeter is connected in parallel to ensure that most of the current keeps flowing through the circuit, and only the smallest proportion of that current flows through the voltmeter.

Voltmeters are not 100% reliable, and there will always be a slight error in your readings, especially as you add more components to the circuit. However, if properly calibrated, these errors will prove insignificant in the end.

Measuring resistance using an ohmmeter

The resistance of a circuit is measured using Ohm's law, meaning that an ohmmeter is basically just a combined ammeter and voltmeter. If you know the voltage of the circuit, then the internal battery of the ohmmeter and the current travelling through it can be measured. The meter then divides the battery's voltage by the current flowing through it and gives the answer in ohms.

One of the most important things to remember when using an ohmmeter is to keep the resistance of the leads in mind when doing the experiment. The resistance in the leads can be measured by touching the leads together and taking down the value that the meter returns. You can then adjust your final values appropriately, as you are doing your calculations.

Most meters on sale today have a built in 'null' value, which can be helpful. It works by reading the resistance of the leads and automatically recalibrating itself to consider the value, leaving you free to carry out your experiment.

CHAPTER 5: TYPES OF CURRENT

Before experimenting with circuits, it is important that you become familiar with the different types of electrical current, and what makes them differ from each other. There are two types of current, alternating current and direct current.

Direct Current (DC)
The term direct current(DC) is used when referring to electrical systems that use only one polarity of voltage or current. DC is produced by various sources, including batteries, thermocouples, solar cells, power supplies, fuel cells and dynamos. Direct current flows through conductors as usual, but it can also flow through semi-conductors, insulators and can even flow through a vacuum.

The voltage across a direct current source is always constant, as is the current through the source. As much as DC stands for direct current, it can also stand for constant polarity. When described like this, the voltages produced by DC can vary over time as is seen when measuring the voltage across a telephone line when a voice signal is passing through it.

DC is ordinarily found in extra-low voltage and low voltage appliances. This is especially true for appliances that use batteries or solar power. It is also found in cars, which normally run on a 12V DC current. However, there are cars like the classic VW beetle that run on lower voltages (6V in the case of the beetle) and some that run on higher voltages such as some semis that run on a 24-volt system (2 12 volt batteries connected in series)

Alternating Current (AC)
The term alternating current (AC) is used when referring to an electric current where the charge reverses direction regularly. Depending on the country you are in, AC power comes in two different frequencies, either 50 Hz or 60 Hz.

AC is produced using a generator called an alternator that turns mechanical energy into electrical energy. Alternators work by spinning a copper coil of wire very fast in a magnetic field. The interaction between the spinning coil and the magnet inevitably produces AC power. AC power is produced due to the fact that as the wire spins, it interacts with a different magnetic polarity periodically. The spinning of the wire is induced by various means, including water (hence hydroelectricity), steam (to form hydrothermal electricity), and wind turbines.

AC comes in a variety of forms, and if you were to connect an oscilloscope to an AC circuit and plot its voltage over time, you would most likely see a variety of different waveforms. The most common form of AC is the sine wave, though it can also come in the form of a square wave or a triangle wave, as shown below:

The power that is supplied to homes and businesses arrives in AC form, and is usually transported along the power lines at high voltage (+110kV) to reduce the amount of power lost during transmission due to resistance in the wire. AC is also used to power electric motors in dishwashers, refrigerators and other appliances. The motors in this case are almost identical to the alternators that help produce the electricity. However, in this case, the motors turn electrical energy into mechanical energy, the exact opposite of what alternators do.

CHAPTER 6: INTRODUCTION TO CIRCUIT BOARDS – THE BREADBOARD

Circuit boards come in all shapes and sizes, and for a whole variety of purposes. They exist in almost everything electronic that we use in our lives, from our fridges to our cars. Now that you are familiar with the basics of electronic principles, it is now time to look at the different sort of circuit boards there are and their different uses.

There are many different reasons why you should select the type of circuit board that you are going to use carefully, not least of which is that the type of circuit board you choose will also affect the type of work that you do. There are circuit boards for beginners who look at experimenting with circuits as just a hobby, and there are those for the more advanced hobbyist and the professional. Listed below are the three most common types of circuit boards and their characteristics and uses.

Breadboard

The simplest circuit board there is a breadboard circuit board. It is usually utilized for testing and for building prototypes before the final circuit is built. When people started experimenting with circuits, especially in the early days of radio, many amateur electricians would test their circuits out by nailing or tacking their wires and terminal strips down to wooded boards that were used to slice bread on.

As time passed though, the breadboards evolved slowly until in 1971 Ronald J. Portugal designed the first solderless breadboard, which made it much easier to change connections and replace certain components. The lack of a need to solder your components to the circuit board also meant that not only could you use your circuit board again, but you could also use the components with which you were experimenting.

Modern breadboards are made up of a perforated block of plastic with a number of tin plated phosphor bronze or nickel silver alloy spring clips under the perforations (called sockets or holes). The sockets on a breadboard are usually spaced out at 0.1 in intervals, and the leads for most components of the circuit can be pushed straight into the sockets. It is important to note that stranded wire is not suitable for use with breadboards as they tend to crumple when inserted into a socket, and strands have been known to break up, causing serious damage to the board.

Integrated circuits are inserted across the gap in the middle of the board, with their notch or dot on the left, while all the other components of the circuit (other than the power supply) are inserted into the remaining free sockets.

The sockets in the middle of the breadboard are called the terminal strips. Terminal strips have a notch in the middle which is meant to aid the cooling of ICs that are connected to the circuit (hence the reason why ICs are inserted in the middle of the breadboard.) Terminal strips are connected vertically across the board, usually in groups of 5 and are not linked across the center of the board.

The sockets on the top and bottom of the breadboard are known as the Bus strips, and these are usually reserved to supply the power to the circuit. They are linked horizontally down the circuit board, and usually one of the strips is meant for ground, while the other one is meant for the supply voltage. Normally, the column that supplies power would be represented by red lines when being illustrated in a diagram, while black or blue lines would represent the grounded side. Below is a picture of a simple Breadboard.

It is important to note that on larger breadboards, there may be a gap halfway along the top and bottom supply rows. If you have a breadboard that has this problem, you should bridge the gap before you start trying to build a circuit, as you may forget and leave half your circuit without power. In addition, if you would like to have a larger breadboard, many companies make breadboards that can be stuck together to make them bigger.

Breadboards have quite a few reasons why they are used primarily for testing and prototypes. One of the major reasons is that they have a very large stray capacitance, meaning that they have a tendency to store unwanted charge. Though it can be ignored most of the time, especially with smaller, lower frequency circuits, when it comes to high power or high frequency circuits it can act as a feedback path. This would make the circuit oscillate at the same frequency, rendering most circuits unusable.

Breadboards also have high inductance in some connections. This means that in some places on the board, the board itself will be able to generate a magnetic field that will then affect the rest of the components connected to the board. The resistance in these boards is also quite high, and cannot be reproduced easily. They also have very poor voltage and current ratings. For these reasons, breadboards are usually restricted to very low frequencies, usually less than 10 MHz, depending on the size and complexity of the circuit.

Solderless breadboards have some other limitations. For instance, you cannot connect any surface-mount technology devices on the board unless they are compatible with the 0.1 in grid spacing. What is more, breadboards cannot accommodate any devices that have multiple rows of connectors if they are not compatible with the layout of the board.

Complex circuits are virtually impossible to create on solderless circuits due to all the wiring that is needed, and the fact that there is no concrete way to secure the wires. The fact that the breadboard is solderless means that it would be too easy to disturb the wires, meaning you would probably spend 90% of your time checking connections, rather than making sure that the circuit works properly.

Creating a Circuit on a Breadboard

When creating a circuit on a circuit board the first thing that you need to do is create a circuit diagram. However, converting the diagram to the breadboard layout can be tricky as the way you arrange the components on the board will look very different from the circuit diagram.

When placing the different components of your circuit on the breadboard, the most important thing to concentrate on will be the connections on the circuit board and not necessarily their positions on the diagram. The best place to start would be with the IC chip. Once that has been connected, you can begin to work around it slowly, pin by pin, making sure that you have connected all the components to the pins they are supposed to be connected to.

The best way to explain the process is to go through an example. Below is a diagram for a monostable circuit that is rigged to light an LED light for about 5 seconds when the trigger button is pressed. The time that the light will be on is determined by R1 and C1 (where C1 is a capacitor), therefore should you wish to change how long the light comes on you should experiment by changing their values. The value for R1 should be between 1 kΩ and 1 MΩ.

Time period can be calculated by:
Time period $T = 1.1 \times R1 \times C1$

To build this circuit
1. Insert the 555 IC in the centre of the breadboard and ensure that its notch or dot is on the left
2. Connect a black wire to 0V
3. Connect the 10 kΩ resistor to the +9V power supply. Connect a

push switch to 0V (remember to solder the leads onto the switch)

4. Connect the 47 kΩ resistor to a used block of 5 holes, then connect an LED from that block to 0V

5. Connect a red wire to the +9V row of sockets

6. Connect the 0.01 μF capacitor to 0V. If you find that the leads are too short, take a wire and connect it to an empty block of holes, then connect the capacitor to that.

7. Connect the 100 μF capacitor to 0V and connect the lead to pin 6, then connect a blue wire to pin 7

8. Connect the 47 kΩ resistor to +9V. Make sure that there is a wire already connected to pin 6

9. Connect a red wire to +9V

When you have done all this:

- Check all the components are connected properly
- Make sure that all the parts are connected the right way around, especially the LED and the 100 μF capacitor
- Ensure that no leads are touching unless they connect to the same block
- Connect the breadboard to a 9V supply and press the switch to test the circuit.

If the circuit fails to work, disconnect the power supply and re-check every connection against the circuit diagram. Once you have the circuit working properly, remember you can experiment with the circuit by changing the capacitors and the resistors to see how long the LED stays on once you have pressed the switch.

CHAPTER 7: STRIPBOARD

Stripboard is the next level up from breadboard, and just like its simpler counterpart it has perforated holes or sockets that are arranged in a regular grid, with 0.1 inches separating the sockets. Also similar to the breadboard is the parallel strips of copper that run underneath the stripboard from one side of the board to the other and are separated by 0.1 inches.

Unlike breadboard though, stripboard is rather permanent, as components are usually soldered to the board. This makes it ideal for small, simple circuits with just one or two integrated circuits. However, the large number of holes on the board means that it is very easy to connect components to the wrong sockets, therefore more care is needed when working with stripboard than with breadboard. For larger, more complex circuits, it is recommended that you use a printed circuit board, as it will be much easier to track the connections that you make.

Stripboard first appeared on the market in the middle of the last century, and was created to be an alternative to the printed circuit boards that had become common at the time. The printed circuit boards at that time had the components that were connected to them placed in a regular pattern.

The stripboard was developed so that you could be free to connect different components as you pleased, as long as they were connected to the conductors in the right way to be able to complete the circuit.

Stripboard, like breadboard, does not need any special preparations in order to be used, save cutting it to size. It is delivered in various sized, with one of the more common sized being 160 mm x 100 mm. However, these dimensions may prove to be too much, hence the reason why you may need to cut it to size.

Stripboard can be cut with a junior hacksaw, or (for those who trust their aim) it can be simply snapped along the lines by placing it on the edge of a table and pressing down hard. If you cut the board in this manner, please make sure that you use a pair of pliers to get rid of any jagged edges.

You may need to make additional preparations if you would like to make a more complicated circuit. This preparation comes in the form of cutting the tracks on the stripboard. In fact, most circuits that are built on stripboard have had to have some of the tracks cut so to break the connection at that point.

If you would like to connect an integrated circuit to the stripboard then you are going to have to cut a few tracks, unless it is one of those rare cases where it is alright to connect opposite pins. Tracks are usually cut with a 3mm drill bit or a special track cutter tool. The tracks are cut from the bottom or the copper side, and the position that the cuts need to be made are usually marked with a cross (X). However, the nature of the board means that even with the marking it is very tricky to find the right place to cut. Should you ever have this problem, it is recommended that you solder your component onto the board before you cut the track. The solder joint will make it easier to identify the right track to cut.

When cutting the track, place the track cutter (or drill bit) on the correct hole and twist it to and fro, making sure not to apply too much force, after all, the aim of this exercise is to cut the track, not punch a hole through the stripboard. Once you have made the cut, make sure you thoroughly inspect your work to ensure that you have no stray copper wires left across the break, as even the smallest sliver will conduct electricity.

Connecting Components to a Stripboard

When connecting components onto a stripboard, it is important to remember that they are placed on the non-copper side of the board. Once they are placed properly, the board is then turned around and the leads are soldered onto the tracks as shown below.

Most illustrations of stripboard layouts display the board from above so you never get to see the tracks underneath the board. More often than not, they are shown with the tracks running horizontally across the diagram. The nature of the stripboard means that the components need to be connected very carefully. Confusion brought on by the large number of holes is not uncommon, and many people have had to restart projects because of avoidable errors. The best way to steer clear of these unfortunate events is to solder the integrated circuit onto the circuit board first, then position all the other components around it.

The position of the components as far as the horizontal is concerned is not really an issue, as regardless of whether the component is too far right or left, it will still be connected to the right tracks. However, should a component move too far up or down then this could be a problem as it will connect to the wrong track and therefore the wrong part of the circuit. There are some people (and more recently manufacturers) who create a grid reference by labeling the holes with letters and numbers. However, if it's being done manually this still takes a lot of work, especially if you want to make a complex circuit.

Connecting external wires to the circuit board is quite straightforward, and is basically the same as connecting components to the board normally. All you need to do is solder the wires through the holes. On the other hand, should you have wires that are too thick to fit through the holes in the stripboard, there are special pins called Veropins which fit snugly into the holes and can be soldered onto the external components, making it easier to connect them to the board.

Planning a stripboard layout

As with breadboards, the first step to take when you want to build a circuit is to draw a circuit diagram. Once you have done that you can convert the circuit diagram into a stripboard layout. However, just like with the breadboard, the circuit diagram and the stripboard layout will probably look like two different things. This is because as with the breadboard layout, what you need to do is concentrate on the connections between the components, not the appearance of the actual circuit.

Before you start planning the circuit, you need to collect all the parts you will be using so you can work out how much space they will require. For instance, most resistors will need at least 3 hole of space to be if they are going to lie flat, but it should be noted that they can take a lot more space. They can also take very little space if you install them vertically. However, this makes the tracks and indeed the resistor more vulnerable to damage. For this reason, if you are thinking of having the stripboard layout last for a long time, it is recommended that you mount all the resistors horizontally.

Planning the layout can be done in many ways. You can be traditional and do it with a pencil and paper, or you could design it on a computer if you have the software. It is important to check with your circuit diagram before you solder any part of the circuit, as the diagram will help you should you not have connected anything properly.

Downloading a stripboard planning sheet is a sure fire way to make sure that you can make your plans as accurate as possible. These sheets have a 0.1-inch grid marked on them to match the spacing on the board. Working with a full scale 'model' of the circuit will allow you to make your connections as accurate as possible and test how much space you are going to need for certain components.

Below is a step-by-step example of how to plan for a 555 astable circuit which flashes an LED.

Please note: like in the first experiment, the LED flashes at a rate that is dependent on the resistance of R1 and R2, and the capacitor C1. The resistance that R1 supplies should be no more than 1 kΩ and combined, the resistance of R1 and R2 must not exceed 1 MΩ. Below are the equations that you may use to calculate the length of time the LED lights up.

LED on time: $Tm = 0.7 \times (R1 + R2) \times C1$
LED off time: $Ts = 0.7 \times R2 \times C1$
$T = Tm + Ts = 0.7 \times (R1 + 2R2) \times C1$
Frequency (flashes per second): $f = 1/T$
Tm and Ts are basically the same if R2 is significantly larger than R1

Planning the Layout
1.   Place the chip holder in the middle of the planning sheet, ensuring that pin 1 is on the left.

2.   Mark the places where you cut the tracks using a cross (X). This is to stop opposite pins from being connected together.
3.   Choose the tracks that will connect to the power supply. Make sure that they are at least two or three spaces above the last track and below the chip holder, like in the diagram above.
4.   Add the wire links by drawing a blob (•) at each end. The links are drawn vertically because the tracks below the stripboard make horizontal connections. Remember to work around the chip from pin one, and draw all the links to the supply tracks. In the diagram, pin 1 is connected to both +V and 0V, while pins 4 and 8 connect to +V.

Make sure that you draw any links that are needed between pins on the same side. There are none in this particular example but they are relatively straightforward to add.

Links to pins on the other side of the chip should be handled with care. In the example above, track 2 and 6 are connected because there are two links that are connected to a track that is below the chip. This is not the only way to connect pins on the other side of the chip. If the points are across from each other on the same track, all it would take would be a repairing the broken track between them. The other way to connect two pins that are on opposite sides of the chip is to curl an insulated wire around the chip.

5. This is the most confusing step, as this is when you will start adding components that will be added to the board. This is when you will also be able to gauge how much space you will need for the components. It is an important step because it tells you how much space you will need, and perhaps how large your circuit will be. This is the hardest stage in the planning process and your plans will change numerous times at this stage before you finally settle on a layout.

This is also where you can find alternative connections to work with. For instance, if you have an LED that needs to be connected to 0V, but the link is too short, perhaps you could connect the link to the track with the first pin, as this track is already connected to 0V. Notice that connections that do not need the IC are made using an unused track. For instance, if you look carefully you will see that the resistor R3 is connected to the LED via a free track above the IC.

6. Add wires that will enable you to connect components that are not on the stripboard such as switches. These are usually placed at the extremities of the board. You should start off with the batteries or power supply leads. The other connections are easier as they are all off the board and therefore you do not need to worry about their size.

7. Check your plan carefully to make sure that all the connections have been done properly, and that every connection that is in the circuit diagram has been made on the circuit board. One way to do this is to go around pin by pin just the way you did in the beginning, making sure that all the components that are meant to be connected to the circuit are connected.

8. Look for ways to improve your plan, as most times there is a way that you can make the circuit more efficient. For instance, in this case, once all the components have been connected, the circuit can be made slightly smaller by moving the +V and 0V tracks as shown below.

9. After you have done this, check your plan again, and this time, make a neat copy. Make sure you have labeled everything clearly, including all the references and values. Calculate the size of the final circuit board and make sure you leave a hole on the left and the right of the board. These holes are left so as to avoid soldering at the end of the track.

As much as you may want to get straight into the soldering work, you will still need to check your plan very carefully first to make sure that you have made no mistakes. Remember, as this is a stripboard, you will not be able to correct mistakes that easily if you get anything wrong while you are soldering.

CHAPTER 8: PRINTED CIRCUIT BOARD

Printed circuit boards (PCBs) are circuit boards that contain copper tracks connecting the holes where the components are placed. Each PCB is unique, as it is created for just one circuit, making the construction and assembly of the circuit very easy.

The copper circuits are printed onto an insulating board called the substrate. The different components of the circuit are then soldered onto the different circuits. A printed circuit board may have circuits on it that are designed to do just one thing, such as turn on a light, while others are designed to perform a whole range of functions.

Printed circuit boards come in three types. There are single sided PCBs, where there is only one layer of copper in the circuit board. Single sided boards are designed to hold simple circuits. Once there are too many components for a single sided PCB to handle the double sided board comes in to play.

Double-sided PCBs have two copper layers in the board. Connections between the circuits on each side of the board are made by plating the insides of strategically drilled holes with a conducting material. Although double sided printed circuit boards allow for even more complex circuits, the most complex circuits are made using multilayer PCBs.

In multilayer circuits, the substrate has layers of conducting material embedded into it, and separated by insulation. To connect components to the different conducting layers, holes are drilled to the appropriate level and plated a conducting material. The components that need to connect to the different layers do so by connecting to the appropriate hole. This usually leads to a much simpler, neater circuit board.

Components on a printed circuit board are connected in two main ways:
- Through Hole Technology

Through hole connections are made just as the name suggests. The components all have small wires or leads that are pushed through small holes in the circuit board. These leads are then soldered to the connection pads on the opposite side of the board.

- Surface Mount Technology

With surface mount technology, J or L shaped legs on the components are soldered directly on to the circuit board. This is done by first using a mixture of solder, glue, and flux to hold the board together at the point of contact, usually until the solder has melted and the connection has been finalized. However, you could wait for a few hours as the solder is "reflowed" in an oven to complete the connection.

Surface mount technology needs a lot more concentration to carry out properly, but it does get rid of the need to drill holes into the substrate.

Parts of a circuit board

Circuit boards have different parts which work together for a complete connection. These are:

i) The board- the board of a circuit board can be made of different materials. There are the older boards, which were made of paper and phenolic resin and there are the newer ones which come in various color, made from glass fiber and epoxy. The materials used in the making of circuit boards have to be safe from any chemicals; electricity and any kind of thermal stress for as long as it will e in use. You must also pick a material that will be easy to cut ad drill if you want to create your own circuit board.

ii) The traces- these form the wiring of the circuit board. Components will be soldered on the traces in order to form electric circuits. They too come in different sizes and shapes. In order to create the traces, a copper foil is used; it is attached to one side of the board, then printed with a pattern. The exposed parts are them fixed chemically in order to leave lines that are conductive and these are the ones called traces.

iii) The components- these are the various pieces that are soldered on both sides of the circuit board. They come in different shapes, and the shape of a component will determine how it will be fixed on the board. Different components have different electrical requirements too.

iv) The connectors- these are the attachments that supply electrical signals and power to the circuit board.

v) The vias- some circuit boards have layers of copper on both sides of the board, and this copper has to be connected at one point. A hole is drilled through the board and a conducting metal, which connects the various layers is filled on that hole. The vias are these holes. A circuit board can have so many layers of copper and as many vias as is necessary as this is what increase the density of a circuit board.

Materials Needed to Create a Printed Circuit Board

The most common materials used to create printed circuit boards are a fiberglass epoxy resin for the substrate, with a copper foil glued to one or both of the sides. There are also PCBs made with a paper reinforced phenolic resin instead of fiberglass. These PCBs are cheaper than their fiberglass counterparts and because of this they are used widely in household electrical devices.

The circuits themselves are made from copper. To draw the circuits on the board, the copper is either plated or etched onto it until it has left the desired pattern. A layer of tin-lead is then used on the copper to prevent oxidation of the metal. The contact fingers on the side of the board are not only coated with the tin-lead compound, but they are also coated with nickel and gold to aid in conductivity.

Processing and Assembling a Printed Circuit Board

The manufacture of a printed circuit board is usually carried out in a very clean environment that is as free of contamination that you can get. Most manufacturers have their own unique processes but the following process is what would be used to make a standard double sided printed circuit board.

Creating the substrate

1. Fiberglass is fed through a processing station where it is impregnated with epoxy resin by dipping or spraying. Rollers then roll the fiberglass to the desired thickness for the final resin. Any excess resin is also squeezed out during the rolling process.

2. The substrate is then passed through an oven to cure it, and then it is cut into large panels.

3. The panels are then stacked up in layers alternating with layers of copper foil. The stacks are then placed in another oven and subjected to high temperatures and pressures for about an hour to fully cure the resin and ensure that the copper is tightly bonded to the surface of the substrate.

4. The panels are then stacked up on each other and secured together to make sure that they cannot move. They are then introduced to a CNC machine where holes are drilled into them depending on the pattern that was entered into the machine when the boards were laid out.

5. The holes that were intended to hold conducting material are plated with copper, while the ones that were meant to be non-conductive are plugged to keep them from being plated.

Creating the printed circuit pattern on the substrate

Two processes are used to print the circuit pattern onto the substrate, one is additive and the other is subtractive. In the additive process, copper is plated onto the surface of the substrate, and the rest of the substrate is left unplated. In the subtractive process, the entire substrate is first plated, then the areas that are not wanted are etched away, leaving the circuit in place. The following section will describe the additive process.

6. The foil surface of the substrate is washed to rid it of grease. The panels then pass through a vacuum where they are coated with a layer of photoresist material. The vacuum ensures that no air bubbles will be trapped between the foil and the photoresist. The desired circuit pattern mask is placed over the photo resist and then exposed to ultraviolet light. The clarity of the mask in the areas of the circuit pattern ensures that the photoresist in those areas becomes exposed to the ultraviolet light, making it more soluble.

7. Once the mask is removed, the panels are sprayed with a solution that dissolves the irradiated photoresist leaving the copper foil exposed.

8. When electroplating the copper, the foil on the copper of the substrate acts as the cathode. The copper that is plated onto the exposed foil has a thickness of about 0.001 – 0.002 inches. The area of the substrate that is not exposed cannot act as a cathode and therefore does not get plated. Tin-lead is then plated on top of the copper to prevent oxidation.

9. The remaining photoresist is exposed to a solvent to help remove it from the rest of the copper foil. The boards are then sprayed with acid to that eats away at the copper. The circuit pattern remains unaffected as it was coated with the protective tin-lead.

10. Contact fingers are then attached to the edge of the substrate, masked off from the rest of the board, then coated with three metals, first tin-lead, then nickel and finally gold.

11. The tin-lead coating on the copper printed circuit oxidizes easily, therefore to protect it, the panels are passed through an oven or hot oil bath again, which causes the metal coating to melt into a shiny surface.

12. Each panel is then sealed with epoxy again to protect the circuit while the components are being attached to it. Various markings and instructions are then penned into the boards

13. The panels are then cut into their respective boards, and then smoothed.

Before the components are added to the board, a test called the bare board test is carried out. This test looks for "shorts" or connections that are there that should not be there, and "opens" or connections that should be there that are not. In processes like the one described above where the PCB manufacturing is on a large scale, the company would probably have a rigid needle adapter that would check the connections on the board. The adapter is a relatively expensive piece of equipment and therefore is unsuitable for small and medium scale producers.

Small and medium scale produces would rather use flying probe testers to ensure that their circuits are functioning properly. These probes are moved over the board by an XY drive, where they are then instructed to apply a voltage to the different contact points on the circuit.

Once this has all been done, the components need to be mounted. This can be done by an individual at home or in the factory. If you are the one who is about to connect components on to a circuit board then you have to keep in mind, just as with the breadboard and the stripboard, the most important thing for you to concentrate on is the connections that you are making, not the way the circuit looks. The good thing about printed circuit boards is that you can use your computer to preview and test out the circuit before you even get the board made.

Software such as Yenka Electronics allows you to design and test your circuit. It allows you to use over 150 different components including resistors, integrated circuits, capacitors, LEDs and switches, to create a circuit on your computer. The completed circuit can then be modified in countless different ways until it works the way you would like it to.

Software like this almost guarantees that the only mistakes you shall be making are when you have to solder the components onto the board. However, that may not be such an issue as there are machines that are designed to make those connections for you, so the only thing you may have to do is design and test the circuit in the virtual world.

Identifying components on a circuit board

Every electronic equipment is made up of various components which work together in order to adjust and run voltage and current in different ways. Most of these components are quite normal, meaning that they can easily be identified by their shape and color and so, they can be bought over the counter. With time, you should be able to identify most of these components and what role they have to play in your circuit board.

1. Use a schematic diagram

A schematic diagram of a circuit board can be the best guide for you to help you identify components on that circuit board. On a diagram, you can clearly see wiring and connections between different components, which is clearly shown by straight solid lines. There are short lines that are usually zigzagged on the schematic diagram too, which represent the resistors. There are capacitors too, which are shown by short lines that run parallel to each other. They can be straight or curved. There are diodes and transistors too, which have arrows showing the direction of the current flow. In a diagram, you will see the complex components too, which in most cases are represented by a block illustration. Schematic diagram is a clear representation of everything you will see in a circuit board, therefore you can learn a lot from it.

2. Integrated circuits

These are also called ICs. They are very small electronic components which contain so many transistors and other parts of components. These parts are usually microscopic. They come in different styles and designs and they connect to a board through several metal pins. Their parts are usually numbered to make it easy for users to spot them with ease.

3.   Use of a circuit board

You can still use a circuit board in order to learn how to identify components in a circuit board. There are different types of circuit boards that you can use for this; there are the large ones and the small ones. A characteristic circuit board has components on one side and conductive copper foil paths on its other side, which is basically the wiring of the circuit board. Circuit boards come in different colors too.

4.   The capacitors

These too have been mentioned above and they are the main storage containers for electric charge. They are cylinder-like, disk shaped and sometimes they look like sleep gumdrops and they can appear vertically or horizontally.

5.   Resistors

These are very simple electronic components which put a limit on the amount of current that is passing through a circuit. Hey have been discussed briefly in a previous chapter. You can easily identify them on a board because they are small cylinders that run horizontally, with maybe four or five stripes that are colored. One single circuit board can have so many of these resistors.

6.   The connectors

These are used to pass electrical signals to other parts of the electronic equipment; therefore, they have to be connected to a cable. A circuit board can have as many of these as possible. Most of them are plastic, with a few metal pins, which connect to the cable.

CHAPTER 9: TESTING THE CIRCUITS

Now that you have an idea of the three types of circuits and their capabilities, and having built at least two circuits by yourself, it is time to find out how to test these circuits. The easiest way to test these circuits and make sure that they are performing at the ideal level is by using a multimeter.

A multimeter is an instrument that was designed specifically to deal with electrical measurements. It can take readings for a whole range of characteristics including electric current, voltage, and resistance. Multimeters are used to test for everything from how much current is flowing through a particular circuit to the potential difference of a household battery.

When being used to test circuits, they are usually used to test the continuity of the circuit, that is, to see if current still flows through the circuit. However, the vast majority of functions that the multimeter has also allows you to check other things that could be wrong with the circuit, such as checking to see if the potential difference of the circuit is optimum, or if there is too much (or too little) resistance in a circuit.

Testing circuits for continuity is rather straightforward, though it differs slightly for the different circuits. For instance, when testing for continuity in a breadboard circuit, the first thing that you need to do is find a pair of jumper wires that fit the breadboard (the breadboard should have come with some) and connect them to both multimeter probes. If the need arises, use alligator clips to connect the jumper wires to the probe. Once the multimeter is connected to the jumpers, set the multimeter to test resistance.

Locate one of the busses on the breadboard that is used for either ground or power connections. Next find a row that has five holes labeled "A" to "E" or "F" to "J". To test the continuity of the "+" column of this breadboard this breadboard, touch one of the multimeter probes to the top hole, and touch the probe to the bottom hole on the same column. If the column has continuity, then the reading that you get will be zero or close to zero. You can test the continuity of the "-" column in the same way.

If you want to check that the two busses are isolated from each other, then all you have to do is touch one probe to any hole in the "+" column and the other probe into any hole in the "-" column. If the result is infinite resistance, then the two busses are completely isolated from each other.

To test a row's continuity, touch one probe to the hole labeled "A" and another to the one labeled "E". If the number is zero or close to zero, then the row has continuity. To check that two different rows are isolated from each other, touch one probe to the hole labeled "C" and another to the hole labeled "C" in the row below. If the multimeter returns an infinite reading, then they are completely isolated from each other.

Performing the continuity test on a stripboard is just as simple. In this case, the only thing you have to remember is not to try and check for continuity along a row where you had to cut the tracks due to an IC.

In printed circuit boards, the continuity of the circuit is usually checked when the board is printed. However, should you want to make sure that you do have continuity and isolation where you need them; you can test the relevant tracks to make sure that they do in fact have these properties.

Digital multimeter

As I mentioned above, a multimeter can be used to measure resistance, current and voltage in electronic circuits. The digital multimeter is also able to test for diodes and continuity. The good thing with these devices is that they are small in size, light in weigt and they can be operated using batteries. These, together with the fact that they can test for a wide range of electronic components makes them highly indispensible for people who want to test and repair their electrical circuits.

How to use it to measure resistant

☐ Connect your multimeter to the circuit. Ensure that the black and the red probes goes into the right terminals

☐ Wind the selector knob in order to set the multimeter to measure resistance. Check for the unit used to measure resistance on your multimeter. In most cases, it will be indicated by the Greek Word Omega, which stands for ohms, which is the unit used to measure resistance.

☐ Turn of the power to the circuit

☐ Take the resistor that you want to measure. Do not leave the resistor in the circuit if you want to get a precise reading

☐ Ensure that the tips of the probes are touching the resistor on each side.

☐ Now read what will be displayed on the multimeter's screen, ensuring that you are taking the correct units.

Use of a digital multimeter to measure voltage

☐ Again, start by connecting your multimeter to a circuit. The black probe will go to the common terminal while the red one goes to the terminal that is marked for measuring ohms and volts.

☐ Set your multimeter for it to measure the voltage that you want to measure. You can for instance choose to measure the direct current, which is the volts DC; you can also measure the millivolts DC or your volts AC, which is the alternating current. If you are using a multimeter with an auto range function, choosing the above settings will not be necessary.

☐ Place the probes across the component in order to measure the AC voltage. Do not observe polarity at this point; do this when you are measuring millivoltage or DC voltage. To measure these, the black probe will go to the negative side of the component while the red one will go to the positive side.

☐ Now read the display, ensuring that you are observing the units as well. The digital multimeter has a touch-hold feature which can be used to keep the reading on the display even after you remove the probes. This way, you will not forget your reading easily.

Use of digital multimeter to measure the current

☐ To measure the current of your circuit, you will either use the terminal that is marked 10 amps or the one that is meant to measure 300 milliamps. Use the former if you are not sure about how much current the circuit has but if you are sure that it is less than 300 mA, you will use the latter terminal.

☐ Change the settings of the multimeter to be able to read the current. In most multimeters, the part that you should be looking for is clearly marked with an A.

☐ Switch the power to the circuit off.

☐ Break the circuit and then place the multimeter in succession with the circuit. The probes will go to either side of the break. You should observe polarity at this point, where the black probe goes to the negative side and the red one to the positive side.

☐ Turn the power on to allow some current through to the circuit. The reason you should observe polarity here is in order to allow the current to go through the current then to the red probe and out through the black one.

☐ Check the display in order to read the results of the measurements. Ensure that you know if you are measuring amps or milliamps. The touch-hold feature can be used in this case as well, to keep the readings on for a while.

How to test diodes

- To get started, slot in the black probe into the common terminal, and the red probe into the terminal that should be used for measuring volts, ohms and for diode testing.
- Choose the function that will be used to test for diodes using the selector knob. Ensure that your arrow is pointing clearly to what you want to test.
- Turn the power to the circuit off.
- Start by testing the forward bias. For this, the red probe will go to the positive side and the black probe to the negative side. The results will be good if they indicate a reading of more than zero but less than one.
- Now test the reverse bias by reversing your probes, to have the red probe on the negative and the black one o the positive. If you get a result of an overload, indicated as OL, you should know that the bias is goods on that side too.
- If the results come as completely opposite, for instance an OL or 0 when you are testing for the forward boas and a 0 when you are testing the reverse boas, you should automatically know that the diode is bad.
- Some digital multimeters beep when the reading is good and also when it is bad

Multimeters are very important to anyone that is interested in electronic circuits. They can be used to measure switches too, but it all depends on the kind of switch. Use your multimeter to measure your ordinary mechanical switches and you will be able to know if they are working as they should or not.

Multimeters cann0t be used to test conditions that could represent a major shock risk for instance when a high voltage circuit is dead or even when its battery is about to die.

CHAPTER 10: FINDING A FAULT IN A CIRCUIT BOARD

A lot of problems in circuit boards are very obvious, even to the people who are not electricians or engineers. There are common problems and simple ones which we all should be aware of because they will make computers, systems and other gadgets not to work as they should. Good thing is that it is easy to find a fault in a circuit board if you take some time to inspect it visually. This way, you will not have to go to a repair shop for something that you can easily fix at home. Knowing how circuit boards work is a skill that will help you find faults easily. Take care not to get shocked though by turning off power before you start to work. Here is a step by step guide to follow in finding a fault in an electronic circuit:

☐ If you are dealing with an electronic or a computer, you will have to remove its cover carefully. Sometimes these covers are held to the device by screws. Carefully remove them, and then lift the cover off.

☐ Start by checking if all your circuit boards have been tightly plugged in, whether you have just one or many. Sometimes the circuit boards get loose when you keep moving them especially those which have more than one pin edge connectors. To be sure they are all okay; push the cards in into the slots which they connect with. This way, there will be no doubt that they are all connected well.

☐ If the circuit board is still not working, you might have to check the plug-in components one at a time. Some of the chips, which fit into their own sockets for instance the daughter boards and the piggy back, could have a loose connection. Remove them and check if they are dirty or rusty as this could be the problem. Once you are done, push the back in carefully to ensure that they are properly in place.

☐ If still your circuit board is not working, check to ensure that there is no water or other foreign objects anywhere on your circuit board as these could be the cause of the problem. Something very small on your circuit board could be the reason it is not working.

- If that does not work, you will have to check all your secondary wiring plug-ins. Some of them could have cable connections that are pushed in for connection and while moving, these could get loose, causing a fault on the circuit board. Once again, pull the off and look for any issue they might have, maybe rusting or a tear. Push them back in to establish a secure connecting and then test whether it is working.
- Check for broken leads on the components too. Some of the components on circuit boards have very small wire leads which can easily be broken close to the component or close to the circuit board. Work on the large and small components to check whether there might be broken wires.
- The other thing you should look for are blackened components of your circuit board, some broken parts, melted parts, the metal lines that join components on both sides of the board and anything else that you might see. This is because anything can happen to your components; they can for instance burn and when this happens, there is usually some smoke. Look for other signs like discoloration of the components or those that might be swollen.
- Finally, check whether there are cracks on the circuit board. Some cracks may look harmless but some of them may have interfered with the circuit traces, causing a problem on your circuit board.

Repairing a faulty circuit board

The very first thing you do when you are faced with a faulty circuit board is to find out what the problem is and the parts of the circuit board that have been affected. Sometimes you might need to repair only a few components and other times some components might need to be replaced with better working ones. Repairing a faulty circuit board will be easy if you know exactly where the problem is. For a common issue that can be fixed at home, here is a guide on how you can go about it:

- First of all, you will need a soldering iron. Turn it on in order to heat it up.
- Now use it in order to remove solder at the pins of the faulty components that require a replacement. The soldering iron has to be hot in order to melt all the solder, which makes the removal of the faulty components easy. Use tweezers to pull out components that are faulty. This has to be done as gently as possible in order not to cause damage on the board
- Once all the faulty components have been removed, use alcohol swabs on the surface of the board to clean up all the pads and the space that surround them.
- Now place the new component in place on the board and its pins

as well. If the component is a through-hole, ensure that you push all its leads through the holes on the surface of the board.

☐ Use your soldering iron to solder the pins one by one in place on the component. If it is a through-hole component, you will have to solder the leads in place on the side of the board.

Note that solder must be used conservatively. This is because too much of it can cause an overflow, which will create shorts if it flows to the nearby pads. You have to be careful about using a soldering iron as well, because its tip can get very hot, causing great burns if it touches your skin.

CHAPTER 11: MISTAKES TO AVOID AS AN ELECTRONICS BEGINNER

Handling electronic circuits is something that you have to master a few times before you perfect your skill in it. When you are handling Electronic circuits, there are so many things that could go wrong, and these are perfected with a lot of practice. There are a lot of mistakes that you can make as a beginner, most of which can be avoided if you take precaution from time to time. Some of the common mistakes that beginners make, which can easily be avoided include:

1. Shorting out things on breadboards

A breadboard is usually a group of rows and columns, which for a beginner may seem as just a small square or nothing significant at all. It is important to understand how things work on a breadboard; otherwise you might end up causing a short circuit especially if you do not know well how the squares are connected to each other. The most important thing in this case is to understand how a breadboard works. This way, you will be able to use it without shorting stuff there.

2. Working with messy fritzing diagrams

You will need fritzing, an open source software project in order to design electronic circuits easily. The program comes freely and it can be used in Windows, Linux and Mac OSX operating systems. Using this software, you can create different types of diagrams, for instance breadboard diagrams schematics and also PCBs. The program is very important especially for a beginner since you can use it to come up with designs that can be explained easily and in case you have a question, you can clearly ask the questions in forums, which can help you to create successful circuits thereafter. However, if the layout is messy the designs create using this tool will not be of any help to you or your audience. No one will be able to understand a messy diagram.

3. Changing connections when you are connected to the power

One rule of the thumb for electricians is never to change connections while they are connected to the power. One of the things that can happen when you do this is to cause a great damage to your circuit. Imagine what would happen if you accidentally put a wire on the wrong connection? What if you touched a naked wire or metal by mistake? There is a lot of damage that can be caused by this. Sometimes you can cause damage and not even know about it, only for the damage components to come to fail later on. If you do not want to deal with the consequences, you will always remember to switch off the power connection before you start changing your connections.

4. Not taking time to learn

There is no way you can be an expert is something if you will not take the time to earn from the basics. Get to understand the important things first, for instance the important terms in electronics and what they actually mean. A lot of people are practicing now and they do not know the difference between volts and power, or even amps and volts. An easier way out is to understand what the terms mean then practice to use them correctly at all times. Think of voltage as pressure for instance and current, which is the one measured in Amps as flow. Finally, take Power as work and you will be able to talk clearly to another person on what you are working on.

5. Not taking devices that are ESD sensitive serious

Electrostatic Discharge, commonly referred to as ESD happens when electric current flow between two objects that are electrically charged. This means that you will not need to establish a direct contact in order for a discharge to occur. When you touch a metal object for instance and you feel some shock, that is how ESD occurs. If it happens when you touch an IC, you will cause damage on it. If the IC or the part of IC that is damaged does not fail to work immediately, it will fail with time. In fact you will not feel the discharge all the time; most of the ESD shocks that cause great damage to components are not strong enough to be felt. The mistake many electricians make is to assume that because they cannot feel the discharge, they will not cause any damage to the component. That is why it is a recommendation for one to use an ESD strap that has an ESD mat whenever you are working with circuits

CONCLUSION

When man discovered how to harness electricity, the world could not even begin to imagine the things that we would be capable of in a century. In the world we live in, it is very hard to go a whole day without interacting with something that has an electrical circuit.

In this book you have covered the basics of electricity, and learned how to use different elements of electricity to your advantage. Understanding how current works in relation to circuits should come in handy one day, especially if you are looking to experiment with your own circuits in the future.

Designing and building your own circuits can prove to be very beneficial. To start with, you will no longer have to deal with professionals for small jobs such as a broken clock or watch. However, creating your own circuits and ultimately equipment is always rewarding.

The different topics we have covered in this book should enable you to dive head first into the world of amateur electronics. By now, should you have followed the examples laid out in the book, you should know how to set up a basic circuit on a breadboard, a stripboard and a printed circuit board.

The ability to create a clock, a timer or a bell for your front door is now in your hands, and these are but a few examples of all the things that you could do with electrical circuits. however, there is no way you are going to know what you are truly capable of unless you start experimenting. So, get out the breadboard or the stripboard, design your first circuit, and get to work. An electrifying life awaits you.

Free Bonus

OneNote User Guide

The Definitive Guide to Learn the Essentials of OneNote in No Time 2nd Edition

Copyright © 2016 Wayne Charles

All rights reserved.

CONTENTS

	Contents	i
1	Introduction	1
2	Chapter 1: OneNote Overview	Pg #
3	Chapter 2: The OneNote Layout	Pg #
4	Chapter 3: Pages, Sections, Notes and Quick Notes	Pg #
5	Chapter 4: Tagging and Searching Notes	Pg #
6	Chapter 5: Advanced Features	Pg #
7	Chapter 6: OneNote and The Internet	Pg #
8	Chapter 7: One Note for Education	Pg #
9	Chapter 8: One Note for Android	Pg #
10	Chapter 9: One Note for iOS Devices	Pg #
13	Conclusion	Pg #

INTRODUCTION

The dawn of the information age spelt doom for the traditional notebook. With hundreds of programs and applications on the market today that will assist people to take down all the information they need, it is no wonder that people have stopped carrying notepads to lectures, seminars and meetings. Everyone from the large corporate conglomerate to the 1st grade student in elementary school is using some form of software to capture all the information they need, and of all the programs that there are on the market, Microsoft's OneNote is one of the best at capturing, manipulating and storing your information.

OneNote has been on the market for over a decade now, and it has been part of the Microsoft Office suite for some time as well. However, it isa program thatis discussed only by people who have experienced it, and its popularity outside the circles of those who use it requires improvement. This is surprising considering just how much you can do with the program. OneNote is more than just a note taker, it is also a planner that is able to capture text, images, video, and audio notes, and retrieve them at the touch of a button. Its compatibility with many of the devices that are on the market today is impressive and means that you can work with it virtually anywhere.

This manual will serve to introduce you to the world of OneNote, and will give you a preview of some of the things that are possible with this powerful program.

CHAPTER 1: ONENOTE OVERVIEW

OneNote is one of the most powerful note-taking softwares on the market today. The program allows users some of the advantages of a word processor, such as the ability to enter text, create tables, and insert pictures. However, unlike most word processors, OneNote users can also add audio and video data to their notes, allowing them to complement their written word with audio/visual data.

OneNote is designed to look like a digital notebook, and in that respect, the designers really got it right. In keeping with the notebook theme, Microsoft OneNote allows users to use virtually ANY part of the page to insert their information, and they can do this by just clicking on an area on the page. Do you want to write outside the margin? You can do that. Do you want to move to the top of the page and write something else? You can do that too. Do you want to just use it like a standard word processor? No one is going to stop you.

The data entry methods for this program are seemingly endless, and that is something that you want from any note taking software in the world we live in, where you never know what format the data you are receiving is going to be.

However, data entry is not the only place where OneNote is beating the competition. For those people who love being organized, this program will be a dream come true as it is insanely easy to arrange your work. The different pages within the program are organized into colored sections, just the way you would have colored tabs within some notebooks, and all these different pages are accessible with just one click of the mouse. Information in OneNote is written into pages, which can be assigned different colors and organized into sections. This collection of pages and sections makes up a notebook, which can be described as the digital equivalent of a tabbed ring binder.

OneNote was originally meant to work on laptops and desktop PCs, but with the passage of time and the evolution of technology, it can now work on virtually any system. Different features have been added to the program over the years to make it easier to work with on tablets and smartphones. This allows users to access the program in places where laptop computers may not be the ideal piece of equipment to use. It also allows users that have stylus enabled smartphones and tablets to literally write information into the program, making it easier to gather information and take down notes when the need arises.

OneNote is also makes gathering information easier by allowing the user to search through and index images and audio files to gain additional information. For instance, if an image has text information that is embedded within the file, OneNote can search for it and find it and display it as text on the screen. It also searches through audio files phonetically, allowing the user to save time when it comes to searching through an audio file. Perhaps one of its most useful features is enabling a user to play an audio file while they are reading notes taken during the recording. This is a very useful feature for anyone to have, however, it is even more useful to people like students and researchers who may need to use this feature to better make sense of the work that they are doing.

OneNote has one of the best multi-user features that you can find in an application today. With this feature, it is possible for anyone with access to make changes to your work. This may seem to be a feature most people would not like to have, until you think about all the professionals and students that have some form of group project that they need to complete. With OneNote, these projects can be simplified tenfold as now all the concerned parties can have access to the information as it is being compiled, rather than having to wait for a meeting where the whole group sits down and makes presentations. The editing of the document can be done at all times, whether the user is online or offline. Notebooks can also be edited simultaneously, allowing more than one user to make changes to the notebook at the same time. This allows users to use OneNote as a sort of digital whiteboard, and also allows users to trade ideas in real time, which allows them to produce the highest quality work that they can in a very short amount of time.

However, perhaps OneNote's best feature is its save feature. Many people who are new to one note have searched the application looking for a save button. However, there is none. Instead, OneNote saves your work onto the OneDrive cloud or a network computer automatically, thereby eliminating the need for you to save every five minutes. This feature is brilliant, as it allows you to concentrate fully on your work, rather than have to remember to save your work at every other turn.

Because OneNote is part of the Microsoft Office Suite, it is compatible with all the other Office Suite programs. You can transfer your work to Word, or Excel and continue it there, especially if you are looking to publish the work that you have done. OneNote was never optimized for publishing works and many of its features actually hint at that fact. For instance, OneNote pages can be ridiculously large, unlike most word processing programs that will give you a specific range of page sizes that you can work with.

Another thing is that there is no set layout or structure to a page on OneNote. This is in stark contrast to word processing software, which always contain some layout or arrangement on a page. Also, a user can load images into a notebook without having to worry about the quality of the photo, as quality is never reduced. Many word processing programs will reduce the quality of the images that are imported into it, usually to save on size and CPU power. However, with OneNote, all the images that are loaded into it retain their original quality. This is unique attribute to have amongst most other word processing and note taking programs, and it is sure to begin a trend that others will have to follow.

As much as there is no specific layout or structure that is prevalent in OneNote, it does come with a whole range of templates that you can choose from. These templates are designed primarily for the user to save time, as within them there are different features already preprogrammed into the notebooks such as to-do lists, calendars and planners, and forms that you can customize.

The program comes with a large number of templates already available for you to use in its library. However, there are also various templates available for download online at the OneNote website and on the Microsoft Office website. Built in templates may be modified to fit your needs, on the other hand, to have a truly personalized experience, you could choose to create your own template design from any of your notebook pages. This is a very useful feature, and it shall be covered in greater detail in subsequent chapters.

Before you get started with OneNote, you will need to open a Microsoft account if you do not have one, so that the program can save your work onto the OneDrive cloud. When you first install OneNote, it will offer you a step-by-step guide on how to do this, and it will only take you a couple of minutes to complete. Once it has completed, you will be able to open the program and begin editing.

In the following chapter, you shall be given an introduction to some of OneNote's basic features, and you will be introduced to some of the changes that were made to the program in the latest version.

CHAPTER 2: THE ONENOTE LAYOUT

Microsoft OneNote has some of the best, most helpful features of all note taking and planning programs. In this section, you shall be introduced to some of the basic features of OneNote, such as how to create a page and a notebook, so that you may better understand the program and be able to begin using it to its full potential.

OneNote Layout

Before we get started on some of the basic tasks that OneNote can carry out, it is important to understand the layout of the OneNote application. OneNote 2016 is very similar in appearance to its predecessor, but there are a few minor changes that you may have to get used to.

At the top left-hand corner of the screen is the Quick Access Toolbar. This is where all the most used commands are housed, such as save, open file and undo features. Directly below that are the Ribbon tabs, which allow you to explore the different tools that are available for you to use in OneNote. One of the changes made to the Ribbon tabs is the addition of a contextual command tab. This tab allows you to select any section of a table or recording to reveal additional features that can be used to modify the table or recording. The ribbon can be hidden from the display by clicking on the pin icon on the extreme right-hand corner of the ribbon display. Clicking the pin icon again brings the ribbon back into view.

In the top right-hand corner is the help icon, which gives users a basic overview of how to use one note, as well as giving you different tips and tricks to make your work easier. Below this is the online login button that allows you to login to the Microsoft servers. Logging into the Microsoft servers allows you to change the settings for your profile, as well as share your work with others so that they may edit or view your notebooks.

On the left-hand side of the screen is the notebooks list column. This displays a list of the recently opened notebooks, and allows you to switch between them at the click of a button. This column also has a pin icon to hide or reveal the contents of the list. The main page dominates the center of the screen like the other Microsoft Office applications. It is in this page that you will enter your information. Information is entered into pages via note containers. These containers can be resized to fit the page by dragging the edges, or they can be moved by left clicking the gripper on the left of the note. Right clicking on the gripper opens a drop down menu that gives you access to additional features.

Above the page on the left hand corner are the different tabs that separate different sections of the notebook. As was mentioned earlier, different tabs can be assigned different colors for ease of access.

The right-hand side of the screen is dominated by the notebook page column. This column allows you to switch between the pages of the notebook you are editing. It also contains an add page button at the top of the column that gives you a shortcut to create a new page. Just above the notebook page column is the search bar that allows you to find anything within the notebook, and all the other notebooks that have been saved in your cloud account, and gives you a simple way to navigate between the pages.

CHAPTER 3: PAGES, SECTIONS, NOTES AND QUICK NOTES

Creating new pages, sections and notes on OneNote is a very easy and straightforward procedure. To create a new page, simply click the (+) Add Page button above the notebook page column. Doing this adds a new page to the currently displayed tab. You could also right click on the desired tab to activate a drop down menu that gives you a number of options, and click on New Page.

To create a new section within the current notebook, click the plus sign on the right of the section tab. You could also click on any tab and select the New Section option. As was mentioned earlier creating a note on a page is as simple as left clicking on any section of a page. You will notice that when you do this, OneNote automatically opens a note container. As mentioned earlier, this container can be resized and moved around the page at your convenience. If you would like to write your notes instead of typing them, select the draw tab. This is also helpful if you would like to create a sketch or draw something into the program.

A Quick Note is basically the digital equivalent of a sticky note, and is unique to OneNote 2016. Unlike notes, sections and pages though, Quick Notes can be created even when OneNote is closed, and be added sorted and edited once OneNote is opened again.

To create a new Quick Note within OneNote, first click on View to display the View Ribbon, then select New Quick Note. Quick Notes can be dragged to any part of the screen, and will remain visible until they are closed, allowing you to make any references that you need to while you work. You can also open a Quick Note by pressing Windows Key (▢) +Alt+N on your keyboard.

Closing a Quick Note does not delete the note as Quick Notes are saved automatically as with regular notes. They are saved in the unfiled notes section of your notebook, and can be accessed by opening your Notebooks list and scrolling to the bottom of the list. To create a Quick Note outside of OneNote, simply press Windows Key+N on your keyboard.

Saving Notebooks in OneNote 2016

Notebooks can be stored on your hard drive or on your OneDrive account. However, it is recommended that you store your projects on your OneDrive account so that you can access your documents anywhere.

If you would like to save on the Cloud in OneNote 2016, you have three options to chose from depending on the type of information you would like to save. Most users will want to save personal information such as assignments, grocery lists, vacation plans and financial information. These can all be stored on the standard Microsoft OneDrive account, and can be accessed only by trusted individuals who have been given the permission to do so.

For those who are using OneNote to manipulate business information there are two options. The first is to use the Microsoft OneDrive for Business platform, which allows businesses to share information on their projects with a small, select, virtual team of people. This makes it easier for companies to organize things like marketing campaigns and product launches.

The second option is to use Microsoft SharePoint Online. This would only be useful for organizations that have teams with existing SharePoint accounts. However, it is a valuable feature for those organizations that have a formal team of professionals that are tasked with duties such as creating schedules or brainstorming ideas.

Organizing your Notebooks

Despite the fact that OneNote is a note taking software, unlike traditional notebooks, OneNote lets you rearrange the sections and pages you have created into an orderly, organized notebook.

Rearranging sections, pages and notebooks is one of the easiest things to do. To rearrange specific sections, pages or notebooks within OneNote, all you have to do is drag whatever it is that you would like to move to its new location on the section bar, page list, or notebook list.

OneNote also allows you to move pages from one section to another, therefore allowing you to fine tune your sections to make sure they make perfect sense. To move a page from one section to another, drag the page tab until the pointer is hovering above the section that you would like to transfer the page to. If you hold the pointer there for a couple of seconds, the section tab will open, allowing you to drag the page to its new location.

Moving a section from one notebook to another follows the same principle. This time however, you are going to drag the section tab to the notebook list column and let it hover above the list until the list opens.

Now, drag the section to the desired notebook and let it go to complete the transfer.

To delete a particular section or page of a notebook, simply right click on the desired section or page and click delete. Should you ever need to recover a section or page that was deleted from a notebook, you can find all your deleted pages and sections in the Notebook Recycle Bin, which can be found in the History tab.

CHAPTER 4: TAGGING AND SEARCHING NOTES

A notebook in **OneNote** can have hundreds, if not thousands of notes. Some of these notes will be extremely important, while others may be insignificant and not really carry any importance to the overall work. Different notes may also be about totally different things. For instance, if you have a notebook that deals with groceries, and you have organized it in such a way that every room in the house has its own section, then that means that the notes that are in the kitchen section are going to be completely different from the section on the bathroom. To differentiate and prioritize the notes in each section, you could tag them. There is no limit to what you can tag in a note and it can be anything from a single line of text to a whole paragraph.

To tag the text in a note, first select the text that you would like to tag. Under the Home tab, the tag button is on the right of the screen. Click on the icon for the tag that you would like to apply to the highlighted text to tag it. For instance if you have text that asks a very significant question that you will need to remember to answer at a later date, you could click on the purple Question icon. If you cannot find the tag that you are looking for, scroll down through the tags gallery using the arrows that appear there.

Please note that the first 9 tags in OneNote have shortcuts going from CTRL+1 to CTRL+9. This is because these are some of the most frequently used tags. Therefore, rather than having to access the tags menu every single time you want to use one of those tags, for instance the To Do tag, you can just use the keyboard shortcut (in this case **CTRL+1**) to tag the highlighted text.

To remove a tag that you have no more use for, click on the home tab and go to the tags gallery once again. Click on the down arrow that appears at the bottom of the box. A drop down menu should open that gives you various options, one of which will be remove tag.

An alternative way to remove a tag would be to highlight the tagged information, right-click it, and select remove tag from the drop-down menu. To remove multiple tags, first select all the text that has tags that you do not need anymore and press **CTRL+0 (Zero)** on your keyboard.

Searching Notes

One of the best features in OneNote is the search feature. It is extremely convenient because it allows you to search all the available notebooks in your archives, not just the one that you are editing. This means that you can start using the program immediately, and not have to worry about where you placed different notes, as OneNote will find them for you instantly.

To search for text within your notes, enter the search keywords in the search box in the right had corner of the screen. Your search results will be displayed in a window that opens below the search bar. Selecting a result will take you to the page which contains the note that has your keywords. You will notice that on the page, all the text that contains your keywords had been highlighted.

You can also search for text that is embedded within pictures, as well as note text. To search for text within images, you need to activate the Text recognition in pictures option. This can be found if you click on File > Options > Advanced. Underneath Text recognition in pictures there shall be a checkbox labeled "Disable Text Recognition in Pictures". Select or clear the checkbox according to your needs.

If you are searching for tagged notes, rather than searching for particular keywords within the note, you can search for the tags themselves. To do this, select the Home tab, and under the Tags group, select Find Tags. A pane will open on the right side of the window labeled Tags Summary. This window will display all the tagged notes in the notebook, allowing you to easily select the note that you need.

OneNote also has the ability to search audio and Video recordings for words as well. However, like with the text recognition in pictures, this feature has to be turned on. To do so, select File > Options > Audio & Video. Depending on what you wish to do, check or clear the "Enable searching audio and video recordings for words" checkbox.

Saving to Other Formats

Like many of Microsoft's products, One Note allows you to save your documents and notes in a different format. This is especially useful if you have a notebook that you would like to share with someone who does not have One Note, or with someone who has an older version of the program. If you would like to save your work in another format, click on File > Save As, then follow the steps that appear. Finally, select the format of the file that is to be saved, for instance, Word 97-2003 document (.doc), and save the document in the selected location.

You can save notebooks as word documents, PDF files, XPS files, or HTML web pages that you can then post online.

These are just some of the basic features, however, OneNote can do a whole lot more than just create notes and search for data within those notes. It is some of these functions that we shall be looking at in the next chapter.

CHAPTER 5: ADVANCED FEATURES

OneNote has some impressive features for those who have gotten a handle of the basics and are ready to really start using the program to its full potential. In this chapter, we shall be looking at some of those features, and how they can help you to improve the quality of your work.

Creating a Section Group by Merging Multiple Sections

While you can group different pages into a section, sometimes it would be beneficial to have sections that are related to each other falling under the same group. This is very helpful, especially when you have a notebook that is so large that the sections tab will no longer display all of the sections in the book on the screen. Should you want to ensure that you can navigate easily and efficiently regardless of the size of your notebook, then this is a brilliant task to learn how to use.

To group the different sections together first open a Notebook that has a minimum of two sections. Once the notebook has opened, right click on any one of the section tabs, and in the drop down menu, select New Section Group. Enter the name of the group and press enter to save it. The group tab should open next to the + sign on the right of all the other section tabs.

Now that you have the group open, adding sections to the group is as simple as a dragging the tab of an existing section to the section group tab to transfer it to that section group. If your section group has multiple levels, you will need to press the green arrow on the right side of the notebook to go back up one level.

Creating a Subpages

Subpages are usually created to help group pages that are closely related to each other. Visually, the only difference between a page and a subpage is that in a subpage, the page's tab is indented, whereas with proper pages, this indentation does not exist. Creating a Subpage is as easy as creating a standard page, however, to create a Subpage there need to be at least two pages open within the book. One page will act as your main page, while the other one will act as your subpage.

To create the Subpage, drag the page that is on the right side of the page tab across the page to the right edge of the screen, until the title becomes indented. Should you want to make the Subpage a main page again, all you have to do is drag the page to the right to make it whole again. However, you could also move the page to the left to indent it even further.

You can hide the contents of a subpage, and indeed the subpage itself, by collapsing one of the main pages. This hides all the levels of subpages that are present under the main page. However, they can also be hidden by clicking on the chevron on the right side of the page tab. This Chevron means that the page has sub pages, and when the chevron is clicked on, lines showing the different subpages that the page contains appear under the page tab.

If you would like to move a page that has subpages under it, just move it like it were a normal page and the subpages will move with it, as long as they are collapsed. However, if you would like to move a subpage alone, you will have to convert it back into a page by dragging it left, and then move the page to its new location.

Obtaining Text from Pictures and Printouts

By far one of the best features on OneNote, the ability to obtain texts from pictures and printouts is extremely useful. The new update of OneNote now supports OCR (Optical Character Recognition), meaning that as long as it can make out the text in a picture or printout, it can read it and copy it into your notes. Once copied, the text can then be edited, which is a wonderful thing for students and people who work in the corporate world who are always getting printouts to read. It is also a great way to get information off flyers and business cards.

It is important to remember that the quality of the picture will determine the accuracy of the OCR tool. If the image is of low quality, or too grainy, it may be interpreted the wrong way by the program. For this reason, it is always advised that you look through the text once it has been grabbed from the image to make sure that it was translated accurately.

To obtain text from an image that you have added into OneNote already, first right click on the picture. Next choose the Copy Text from Picture option, and position the cursor in the area you would like the text to be pasted to.

If you would like to obtain text from a print out with more than one page, you first have to right click on the image of the print out and select one of the following options:

- Either Copy Text from this Page of the Printout or

- Copy Text from All the Pages of the Printout.

In the first option, the program will grab text form only the page or image that is shown or highlighted. However, with the second option, the program will grab text from all the images or pages that are present. Once you have made your selection, position the cursor in the area you would like to paste the information to and right-click.

Recording Audio and Video

Placing and video in One Note is easier than it first appears, and you can even ensure that you know when the audio was recorded by attaching a time stamp to it. One Note will also make the title of the recording searchable so that you can find it easily in the document.

To record audio or video in One Note, first click on the insert tab then select record audio or record video. An audio and video recording tab should appear on the screen, with a timestamped icon giving the date and time that the recording is being made. You must remember that the icon that pops up with the timestamp will be the same as your default audio and video player icon. For instance, if you are using VLC media player, the icon will be the default VLC player striped cone, if you are using iTunes, it will be the iTunes logo.

If you are recording video into one of your notes, a live feed will display in a pop up window as you make the recording so that you can have a preview of what the video will look like. While you are recording, you have the ability to pause or stop the recording by clicking on the pause and stop icons that pop up on the screen.

Once you have completed the recording, you can right click on the video or audio icon and select "rename" to rename the file. This will help you customize your note even more, and add a title to the video that is more relevant to the note that you are composing.

Replaying the video once it has been recorded is also very simple. All you have to do is double click on the audio or video icon and use the audio and video playback tab to control how the video plays. During playback, you can choose to rewind or fast forward the track or video by either 1 second or 10 minutes, depending on the length of the audio or video.

Checking Spelling

One Note allows users to add unlimited amounts of text to different notebooks. This is one reason why it is so important to make sure that all the text has been spelt correctly. Just like other Microsoft applications, One Note comes with an autocorrect feature that fixes common errors in your typing as you type. However, if you would like run a full spell check in the program that is also very possible.

To run a full spell check, first click on the review tab in a particular subpage, and click on the spelling icon. Doing this should open a pane on the right side of the page. Each word identified by the spell checker can be ignored, or changed depending on the suggestions given by the program. Once the spell checking is complete, click OK on the box that pops up and the spelling pane should close.

One Note can also correct any math that you input into it if it notices any common errors. To turn on the Math Autocorrect function, all you have to do is click on file, then options, then proofing, and select Autocorrect Options. Click on the Math Autocorrect tab and check the box next to Replace text as you type.

Converting Handwriting to Text

One of the best features in One Note is its ability to convert If you are one of those people that uses a tablet PC or if you have a tablet attached to your computer, you can draw or write straight into One Note. If you do not have a tablet you can still draw or write using your mouse or trackpad.

To input handwriting or a drawing into One Note, first attach the tablet and pen to your computer if you have them. Next, open a new drawing and select the 0.5 mm black pen (or the thickness of your choice). Write down what you would like the program to interpret for you, and using the Lasso Select tool, select the text you have just written. Then, go to the draw tab in the convert group and click the Ink to Text button. Should any errors arise during the conversion, you can correct them by clicking on the Select and Type button at the top of the page. If you have written down a math problem, rather than click on Ink to Text, you can click on Ink to Math instead, and the information you have entered will be converted into a mathematical expression.

Should you decide to use the default Pen Mode, you will be able to create both handwriting and drawings. However, there are other modes you can use if you would like to create either a drawing or a handwriting entry, such as a Create Drawings Only mode, Create Handwriting Only. You could even use the Use Pen as Pointer mode to use the pen as a pointer during a presentation.

CHAPTER 6: ONENOTE AND THE INTERNET

Sharing Notebooks Online

Sharing notebooks on the cloud is a great way to get others involved in the work you are doing. It is also one of the easiest ways to get work done if you are working as a team as each team member will be able to access the work from wherever they are. It also allows the different team members to work on the same thing simultaneously, and track the changes that each person makes in near real time. The notes are updated every few minutes, allowing all the members of the team up to the minute updates of what all the other members are doing.

If you are editing a note as your team members are viewing it online, they will be able to see all the changes you are making as you make the, and should they change anything, you will be able to see the changes that they make.

Depending on where you saved the document, there are different ways you can shareit to the cloud. For instance, if you saved the notebook on the OneNote desktop program you will need to follow this procedure.

First, open the notebook that you would like to share, and in the menu bar click File > Share. When you do this, you will be given two options, Share with People or Share with meeting.

Should you select Share with meeting, you will have to select Share with Meeting again, and choose a meeting. Alternatively, you could start a new Skype for business meeting and share the notebook there.

If you selected Share with People, you will be supplied with a form where you can enter the names and addresses of the people you would like to share the notebook with. Once you have done that, just click share and the document will be on the cloud.

If you are sharing the notebook from OneDrive, OneDrive for Business or SharePoint Online, then the procedure you follow will be slightly different.

You will begin as you did before by opening the notebook that you would like to share and clicking File > Share. You will then select Share with People, and enter the names and/or email addresses of the people that will be receiving the notebook, and click share.

If you have a notebook that is already on the cloud, you do not need to open it. All you need to do is locate the notebook and click on the ellipsis (...) next to the notebook that you would like to share. When the menu opens, select share and enter the names and/or email addresses of the individuals that shall be receiving the notebook.

If you are sharing to SharePoint Online, you must ensure that all your team members have access to the service, as those without access will not be able to view shared content.

It is also important to note that you cannot share part of a notebook, for instance just a section or a page, you must share the WHOLE NOTEBOOK. However, you are able to set a password to some of the sections of the notebook to restrict access to those sections. If you would like to share just one page or a notebook, click on Home > Email Page. This option will allow you to send a snapshot of whatever you are working on in the program.

Once you have shared the notebook and your team has started to work on it, you and your other team members will have the ability to keep track of the changes that were made by different team members. This is made easy by a feature in OneNote that shows the changes that were made in the document by highlighting the text in bold characters, and placing the initials of the person who made those changes next to them. However, if that is not enough you can get an even more comprehensive look of the changes that were made by using the history tab.

The History Tab

The History tab is used to give the author and members of a team an overview of the changes that were made to a particular notebook, and the identity of the person who made those changes. The different buttons on the history tab include:

1. Next Unread – this moves the user on to the next unread section of the notebook. If this button is not accessible then it means that the notebook has been read in its entirety.

2. Mark as Read – This button is used to narrow down unread content. As the name suggests, it marks sections in the notebook that have been read to show that they have already been analyzed.

3. Recent Edits – This button shows all the modifications that were made to the document over a specific time period

4. Find By Author – This button searches the document for changes made by a specific author.

5. Hide Authors – As the name suggests, when this button is

pressed, the authors of the notebook are either hidden or displayed.

6. Page Versions – This button allows the author to browse through older versions of the selected page.

7. Notebook Recycle Bin – This is where all the deleted notes, sections, and pages go. If you ever need to restore a deleted page, section, or note this is where you will find it.

Taking Notes During an Online Meeting

The fact that OneNote is fully integrated with Microsoft Outlook and Skype for Business means that you can easily take notes in any one of these programs and share them with anyone else.

If you are using Microsoft Outlook for the meeting before you start taking notes you have to open the Outlook calendar and select the meeting that you want to take notes on. Next, open the meeting ribbon and open the Meeting Notes Dialogue box. In this dialogue box you will have two options, one is to share your meeting notes with the rest of the people taking part in the meeting, or to take notes for yourself.

If you would like to share your notes with the rest of the meeting then all you have to do is select Share notes with the meeting. If you want to take your own notes choose Take notes on your own.

A Choose Notes to Share with Meeting dialogue box should appear when you select Take notes on your own. When it does, select a section and a page to record the new notes, then select OK. Microsoft OneNote links all the pages that are used during the meeting to the Outlook Appointment. This means that you are always able to access the notes and other details from that meeting as they are all kept in a central location.

If you do find yourself in a meeting where it is not necessary for you to share the notes that you are recording, you could open the Home tab and choose Meeting Details.

Using OneNote Add-Ins

OneNote has a number of additional programs that you can use to help make your experience with the program better. These programs help you do things like format your content, print documents, capture data from whiteboards and even share team notebooks. Listed below are just some of the popular add-ins that are available for OneNote, and a short description of their different functions within the program.

1. Onetastic

Onetastic is one of the most versatile of all the add-ins that are available for one note. Apart from allowing users to create custom styles for their notebooks, it also displays content in a calendar. Its best feature has to be the Macro functionality, which unlocks a whole new world of possibilities within OneNote. For instance users can now create features like Find and Replace, Daily Journal View and Author Information Removal. There are currently over 180 different macros available on the Onetastic website, and more are bound to be unveiled as people find more ways to simplify their OneNote experience. The best part about it all is that Onetastic and its Macros are absolutely free

2. Office Lens

This is an add-in that is most useful on the mobile and tablet platforms. It allows the camera's on these devices to take all sorts of images, including those of business cards, white boards, and documents, and saves them directly into OneNote. Once the images are saved, they can be loaded into OneNote to have the data in them translated by the OCR.

3. Clip to OneNote

This add-in by OneNote Gem Add-Ins is a brilliant third party application for use with web browsers. It is designed to send the active web page in a web browser as a single image to OneNote's Quick Notes section. It supports all the major browsers, including Internet Explorer, Mozilla Firefox, Google Chrome, Opera, and Safari.

CHAPTER 7: ONE NOTE FOR EDUCATION

One Note is one of the most powerful tools that educators, teachers and students have at their disposal. In fact, it is such a versatile tool that Microsoft themselves created an entire website that is dedicated to One Note and Education, as well special tools that teachers and students can use called the One Note Class Notebook and the One Note Staff Notebook.

These tools allow teachers and students alike to stay organized and work together easier to be able to achieve the goals for the curriculum they are studying or teaching. It also allows teachers to work together to make planning and executing the curriculum easier for them.

The sections that One Note offers allows teachers to section their notes so that they know when different activities should take place, for instance, meetings such as staff meetings, department meetings, and PTA meetings, and all the notes that go along with those meetings. One Note's ability to sync with Microsoft Outlook also allows teachers to have finer control of the notes that you create, and allows them to add meetings to their One Note notebooks easily, including the date, time and people who shall be attending the meeting. If the notebook that the teacher is modifying uses a specific template, the information imported from Microsoft Outlook will automatically adopt the same template, meaning that users will not have to worry about modifying information once it has been imported.

Another thing that teachers will find useful is One Notes ability to gather information from different places while keeping the information from the sources. For instance, teachers who are researching on a particular subject may find interesting websites that they can use in class. The information from these websites can be saved as a screenshot that can then be used in their classroom presentations in class. The screenshot, once imported to one note, will have all the information relating to the source that the teacher needs, including the page URL and information on the website that it was taken from.

There are many teachers out there that also use One Note to organize their to-do lists for different tasks, including administrative and classroom tasks and activities. Once the tasks have been completed you can tick them off just like a normal to-do list, and because they are automatically saved on to the cloud, you can modify them from anywhere, and on any device that has access to the internet.

Important notes and activities can be grouped together using the tags summary, allowing teachers to have easy access to important items, questions and to do lists all in one place. Should the information not have been tagged, the One Note search engine is so powerful that it will be able to find the information they are looking for regardless of how the notebook is organized.

Teachers can also create pages for different students, or groups of students, in each of their classes. By doing this, they can then file all the students information, including contact information, disciplinary records and grades in one place. One Note integration with Microsoft Outlook also makes it easier for teachers to import student information into One Note, and allows them to update information quickly and easily when the need arises. Even when the teachers are busy carrying out other duties, they can use One Note to record their thoughts and ideas by creating Quick Notes.

Creating Fun and Interactive Classes using One Note

Teaching these days has become much more demanding thanks to the increased demand for interesting interactive classes. Teachers can no longer enter a classroom and assume that they will be able to teach effectively using only a text book and a few PowerPoint slides. This is one of the reasons why One Note is such a powerful tool for teachers, it allows them to create interesting, interactive classes that every student will enjoy. The best part of the interactive lessons is that they can be created in such a way that the students will be able to complete them by themselves at a later date.

Lessons can be created with custom audio introductions that help to explain the objectives of the lesson and the expected outcomes of the lesson. One Note's ability to create hyperlinks automatically means that you can attach links to articles, audio, video among other links to the lesson, allowing students to access the additional information that they need easily.

Teachers are also able to link assignments to the lesson. For instance, if you have an assignment that you have prepared on One Note already, all you need to do is link the assignment to the lesson note and students will be able to access and complete the assignment. Once they have completed their homework, One Note allows teachers to grade the assignments in the application, and teachers can even add audio commentary to the assignment to further explain where a student went wrong, or to add praise to work that was well done. An added advantage of doing this is that the comments added can be linked to the audio recordings that you create automatically thanks to the timestamp feature.

Creating assignments on One Note becomes even more important for those lessons that involve practical skills such as music, geography and foreign languages. For instance, if you are a music teacher, you can add audio samples to a particular assignment to test your students' ability to recognize certain tunes or progressions. You could also test your students' knowledge of different geographical features using the different drawing tools provided by the application. Foreign language classes can be made even more interactive by having the teacher create pronunciation assignments that the students complete by following different audio files to test their skills. The teacher could then add their comments of how to improve their pronunciation while they are grading the assignment.

Collaborating with Teachers and Students using the One Note Class Notebook App

Teachers need to collaborate with each other and their students, to ensure that their students gain the best education possible. This is especially important when it comes to class projects, creating assignments, and providing feedback concerning a student's performance. The One Note Class Notebook allows teachers to create notebooks that have preset permissions that have been specially designed to be used in the classroom and in learning environments, whether they are in schools or colleges.

To create a class notebook, you first have to enter Microsoft Office 365, and click on the One Note Class Notebook. You will then be able to create a new Notebook, and enter the name of the class that you are teaching.

Once the class notebook opens, you will realize that it has been partitioned into three different "spaces", these are:

- The collaboration space – this space allows teachers and students alike to edit the information that is in the note. This is especially useful for group work and class projects.

- The content library – this space allows teachers to modify the information in the note, however, it only allows students to view the information and copy the contents if they feel the need to do so. This space is best utilized for giving out tests, assignments and course materials that students may need.

- Student notebooks – these are the students own space where they can work without worrying about other students viewing their work. Teachers and instructors will be able to view every student notebook in their class, and edit the information in the notebook to help improve the student's work. However, no other students will be able to access a private student notebook that is not their own. Every student notebook that you create comes with a few default sections. However, you could modify these sections by adding or removing some of the sections that were suggested.

Adding students to a class is easy, and can be done by manually entering the different students in the class, or by importing a group from Azure Active Directory or Office 365. Should the members of the group change after the notebook has been created, you can modify the group in the One Note Class Notebook app by clicking on Add or Remove Students.

Once you are done editing or creating your class notebook, you are provided with a preview to ensure that you have created a notebook that will benefit yourself and your students. The preview has two different views, one that only the teacher will be able to see, and another that is exclusively for the student.

Once the notebook has been created, teachers are supplied with a link that they can send to the students so that they can access their notebooks. This link can also be used to open the notebook directly, so it would be a good idea to save it somewhere if you are a teacher.

The notebook can then be edited so that it provides all the information that students' will use for that semester. The specially designated sections in the Class Notebook allow teachers to collaborate with students and share information on a convenient platform.

Modifying the teachers or students that are allowed to view the notebook is simple, and only requires the teacher to go to the notebook and click the add or remove students/teachers icons and follow the instructions. It is important to remember that despite the fact that an individual has been removed from the notebook, the information that they contributed to the notebook will remain, though they will no longer be able to access the notebook itself.

Once you have added or removed students or teachers from the notebook, you will be directed to a screen that shows all the modifications you have made. All you need to do to ensure that the changes are permanent is to click on the update button at the bottom of the screen. The app will then create a new link to the notebook that you can send to the new student or teacher if you chose to add individuals to the book.

Any students that use Office 365 will be sent a notification immediately a One Note Class Notebook is shared with them. For those who may not use Office 365, they could also access the Class Notebook by clicking on the "shared with me" folder in Microsoft One Drive.

Sharing Class Notebooks with other teachers is just as simple as sharing them with students, and follows the same basic procedure. You must first add teachers to the notebook by clicking on the "Add or Remove Teachers" icon, then following the steps outlined by the program, which are much the same as when you were creating a list of students. A link that you can send to teachers and other instructors will then be generated.

Perhaps the best thing about the One Note Class Notebook is that you can modify all the notebooks you create from one place. All you have to do is click or tap on the Manage Notebooks icon, and a list of all the Class Notebooks you have created will appear. You can then change the specifics of any one of the notebooks displayed without having to worry about opening a new notebook every time that you need to change a minor detail.

For instance, you can change the name of a teacher or student section, add or remove student sections, and even change the permissions of certain spaces in the notebook. For example, if you feel that your students should not be able to change the information in the collaboration space, all you have to do is click on a button and they will no longer be able to do so. Once you have made all the desired changes, you will have to click on save to update the notebook and ensure that all the changes that you have made are permanent.

Class notebooks can be used to give out and grade assignments. All you have to do is go to the content library and create a new assignment. Once the assignment has been created, students will be able to get the assignments from the content library by copying the assignment into their own notebooks. When the homework has been completed, the teacher can then grade the work by opening the student's notebook and giving comments and feedback on the assignment as described earlier.

The teacher only section group is one of the most useful things for teachers, as it allows them to set up lesson plans and assignments in the same notebook without allowing the students to see what they are doing. It also allows teachers to create content that they do not want the students to access until they feel the time is right.

CHAPTER 8: ONE NOTE FOR ANDROID

One Note may have been designed for use with Windows computers and laptops, but for those who would like to access the program through their mobile devices, Microsoft developed One Note for Android. This application allows users to have most of the features that One Note for Windows has, though it is a lighter, stripped down version of the original. The application was developed mainly due to the extremely small market share that Windows Phones have, and the demand that Android users had for a version of the application that could run on their devices.

For instance, the interface on One Note for Android has been designed specifically for smartphones and tablets, and works just as well as the version created for Windows Phones. The interface if very utilitarian, a fact that becomes especially obvious when you use the app on a tablet.

There are a number of major elements that make up the One Note for Android interface. These include:

1. The Note Pane – this pane takes up the whole screen, except for the top strip that houses the various buttons that you will need when using the app

2. The Up Button – the up arrow button can be found in the upper left corner of the screen, except when you are on the One Note homepage. If you tap it you will automatically be taken back to the section, note or list that you were looking at before you switched to a new page.

3. The Command Buttons – located in the top right corner of the screen, the command buttons are three buttons that work anywhere in One Note except for the home screen. These buttons are usually arranged in the following order:

- Recent Notes: The left-most button of the three, this button gives you access to notes that you have modified or created recently

- New Note: This button is placed in the middle of the three, and allows you to create a new blank note with your cursor in the title field.

- Take a picture: This button is placed on the right of the three buttons, and it allows you to take a photo with your device's camera and add it to your note.

4. Keyboard – immediately you tap on any note so that the cursor appears, the keyboard shall pop up from the bottom of your screen just as it does with any other Android app.

Writing Notes on One Note for Android

Just like with all the other mobile versions of One Note, such as the one for iOS, the Android version of One Note gives you access to your notes, and allows you to edit them (though with much fewer features than the desktop version) at the touch of a button.

To create a new note from anywhere on One Note for Android, all you have to do is tap on the middle button in the upper right corner of your screen. If you are in the middle of viewing a note, the new note that you create will be inserted into the section that you are viewing. However, if you are on the home screen, the new note that is created will be stored in the Unfiled Notes section of your Web or Personal notebook.

It is important that you remember that the icon will not appear at the top of the page if the cursor is in a note, therefore, you may have to tap on the back button on your device to ensure that the button actually appears at the top of your screen.

Opening notes on One Note for Android is just as simple, and only needs you to tap the note's name when you are viewing the section that it is stored in. if you would like to open a note from the home screen, you will first have to tap the name of the notebook that the note is in, then tap the name of the section it is in and finally, tap the name of the note.

List elements are hidden in One Note for Android until you need them. Accessing them is very easy, and involves you opening a note so that the buttons on the screen change. Once you are inside the note, you will see four buttons at the top of the screen. The first button is the camera button, and the next three are list element buttons. They are used to create a numbered, bulleted or check boxed list depending on your preferences.

Recording Audio on One Note For Android

Sometimes the inspiration to record audio for one of your notes will hit you while you are on the move. To help you record all those moments, One Note for Android allows you to record audio, and will even translate the audio you record to text that will be displayed in the note that is open at the time of the recording. You can record into One Note easily using the microphone button included in the keyboard of all Android Devices.

To record audio and have it translated to text, you must first open a new note and tap the area where you would usually type. Once the keyboard appears, tap on the microphone button and speak when the Speak Now prompt appears. You will notice that once you start speaking, the Speak Now prompt changes and becomes Tap to Pause.

Once you have paused the recording, you can choose to do one of two things, ether tap the screen again to continue speaking or tap the keyboard icon so that you can switch back to the standard keyboard and stop recording. As One Note will interpret your speech and translates it into text you may have to correct some errors in interpretation when you are done.

Adding Pictures to One Note for Android

These days nearly all android devices have some sort of camera, and One Note can utilize this camera to add pictures to any of the notes that you are editing. You could add pictures to a note in many different ways including:

1. Using the Take Photo Button: The take photo button is usually a camera icon with a lightning bolt that sits in the upper-right corner of the interface when you are not in an actual note. If you tap this icon, you will be able to takephotos for the note that you are viewing. If you tap the button while you are in any notebook, a note called Unfiled Note will be created and stored in your Web or Personal notebook. Your photo will be stored in this note until you file it in a particular notebook.

2. The Photos Button: The icon for the photos button is usually in the top left corner of your screen when your cursor is in a note. Tap this button to bring up a pop up window that you can use to add photos from your gallery or to take a new photo that will be stored in the note.

Capturing images and storing them in your notes is extremely easy, especially if you are already well versed with the functions of your phone's camera. You must remember that One Note does not have its own camera, and that it only borrows your device's camera to take photos. The functions of your device's camera will be dictated by the version of Android that you are using. However, there are some camera functions that may not be supported by One Note, so you have to make sure that the functions that your camera is capable of are supported by the app before you use them. At the time of writing, One Note for Android still did not support capturing and adding video to the notes that you create on the app.

If you would like to take a new photograph to add in a note page, you can use the following procedure:

1. If you are not already in the note pane, tap it to edit it then tap the camera icon in the upper left corner of the screen. The insert menu will pop up and give you a variety of options that you can use to add images from your gallery or take a photograph

2. Tap capture a photo and use Android's default camera to capture an image. If you do not want to take a photo after all you can always tap the X that appears on the bottom right hand corner of the screen. The bottom left hand corner will usually have a button that allows you to change the settings of your camera

3. Take a photo as you would usually do, the X will remain at the

bottom of the screen, but the camera options button will be replaced by a check mark symbol.

4. Tap the X if you do not like the picture you have taken and keep taking photos until you are satisfied, then hit the check box on the bottom left of the screen. Your note will reappear with the chosen photo attached.

Adding a phot that you have already taken is just as simple, and can be done in the following ways:

1. If you are not already there, tap the note pane to edit it and tap the camera icon at the top of the screen. When the insert menu appears, add a picture from your gallery

2. Tap the add image from gallery icon, and add an image from your selected gallery, be it OneDrive or your phone's gallery. You will then be asked whether you always want to use this source or if you are using it just once, make the appropriate selection and select the photo that you would like to attach to the note. If you tapped always but would like to change the setting later on, the easiest way to do so would be to tap the Clear Data button in the One Note app in Android settings.

Managing Notebooks and Notes on One Note for Android

Despite the fact that it is a more toned down version of its desktop counterpart, you can still manage the different notebooks and notes that you create in One Note on your Android device. For instance, if you would like to delete a note, you can do so by tapping the Delete Page button in the Options Pane while you are viewing the note. However, if you would like to rename, delete or move whole sections or notebooks, you will have to access the OneDrive app or the web app version of One Note.

Settings on One Note for Android

Despite the fact that you have very few settings that you can change in One Note for Android, there are certain settings that you can change that you may find useful. Finding the Options pane on One Note is easy, as the three vertical dots that can be found in the bottom right corner of the screen are more or less universal across most Android apps. However, the buttons do change slightly depending on where you are in the app, and the menu is designed to be context sensitive, meaning that it changes depending on where you are in the app.

For instance, if you are on the Home Screen, or in a Notebook Section, the Options pane will only have three options, Sync, Sync Error, and Settings. However, if you are in a section editing pages, then the Options pane will also have a Create New Page button. Should you be viewing a note, then the Create New Page button will become a Delete Page button instead.

The purpose of the different buttons is as follows:

- Delete page: Deletes the page you are viewing

- New Page: Creates a new page in the section that you are viewing

- Settings: Most of the settings that you can change in One Note for Android can be found when you click this button

- Sync: Clicking this button will sync the current page with the notebook saved on your One Drive cloud storage space

- Sync Error: This only lights up when there has been an error syncing the note with One Drive. Click on the icon when it appears to gather more information on the error

Once you open the Settings menu, you will be able to modify certain things about One Note For Android. Some of the options available to you in this menu include:

- Sync on Wi-Fi only: by tapping this checkbox, you will be confirming that you will only be able to sync notes and notebooks with One Drive when you are connected to Wi-Fi. This will help you to conserve your mobile data.

- Windows Live ID Account: should you ever want to view the account that you are logged in to while using One Drive for Android this is the button you should tap. Usually, you are redirected to a window in the Android settings app, not One Note itself. In the app, there is another options button in the upper right corner of the screen that you can press to either Sync Now, Remove Account or access Help. The Help feature opens a webpage in your browser that allows you to view the help documentation that is related to your device.

- Upgrade: One Note for Android is usually free for the first 500 notes, after which you will not be able to edit or create any more notes until you delete some. However, if you pay a small fee, you can upgrade your One Note for Android app so that you can edit unlimited notes

- Help: Clicking on help opens a webpage in your default web browser that contains all the help information for your particular device.

- Support: Tapping on support will bring up your web browser, and load the different support forums that have been started on the Microsoft website for One Note for Android

- Use Terms: Tapping this will bring up a pop-up window that will display the apps terms of use

- Privacy Statement: Tapping on this will bring up a tab in your default web browser that will display Microsoft's privacy statement.

- Third Party Notice: This item will bring up a pop up window that will give you more information about any third-party technologies that you are using in Microsoft One Note for Android

- Version: Tapping this will give you more information on the version of One Note for Android that you are using

- Copyright: Tapping this option will display the copyright language for the app.

Despite the fact that it is a prominent feature in both One Note for iOS and Desktop versions of One Note, you cannot change the settings for picture quality on One Note for Android. This means that some of your photos may appear to have a lower quality than they actually do, while others will be viewed in their full scale, regardless of what you try.

There are some One Note settings that are not available from within the app itself, but are available through the device settings. To view those settings, you will have to tap the settings icon on your device and select the One Note app in the list of installed apps. Some of the options that are not available via the app but are available in your device's app menu include:

- Force Stop: Tapping this button allows you to stop One Note and close it if it will not close any other way. It stops all processes pertaining to the program.

- Uninstall: Tapping this button will uninstall the app from your device

- Clear Data: Tapping on this button will temporarily clear all the data that is stored on your device by the application. However, when you log in again, the data that had been cleared will be restored.

- Clear Cache: This button will clear all the information that One Note has stored on your system Cache. If the app is misbehaving, you can try to tap this button after you have tapped on Force Stop, then restart the app.

CHAPTER 9: ONE NOTE FOR IOS DEVICES

As the first non-windows devices to receive One Note, the iPad, iPod touch and iPhone were given greater focus, and for that reason, the interfaces on these devices is much better than that on the Android version of the application. In fact, One Note for iOS is even more advanced than One Note for Windows Phone, though it still has fewer functions than One Note for Windows and even the One Note web application.

The One Note for iOS Interface

Navigating through One Note for iOS is a unique experience, as many of the cool features, animations and styles that are used have been optimized to run on Apple devices. All iOS devices, except the iPad have basically the same interface, and the only real difference between the interface on the iPad and other iOS devices is the size of the display, which makes the interface look dramatically different on the iPad.

There are two different orientations used in One Note for iPad, Landscape and Portrait. In Landscape view, you have access to a navigation bar on the left side of the screen, and the note pane is on the right. However, in portrait mode, the navigation pane disappears, and all you have is the note pane that fills the whole screen. The navigation bar is accessed via an icon on the left side of the screen.

On the iPhone and iPod Touch, each pane takes up the whole display, and the buttons are all placed at the top and bottom of the screen. Some of the main features of the interface on the One Note for iOS devices include:

-   The Note Pane: The note pane usually takes up the whole display when in portrait, and two thirds of the screen in Landscape orientation in the iPad version. It looks like a piece of paper in a ring binder on the iPad. For the other iOS devices, the pane takes up the whole display, much like it does on the iPad in portrait orientation, just smaller. Above the note pane you can see where you are in the notebook.To change locations in the notebook or section that you are viewing, all you have to do is tap on the name of the notebook or section that you would like to view and you will be redirected straight to that section or notebook. For instance, if you would like to view a list of pages that are in a particular section, all you have to do is tap on the name of the section. If you would like to view a list of the sections in the notebook, tap on the notebook name and a list of all the sections in the notebook shall be displayed.

- The List Pane: This pane is where you can view a list of your notebooks, the sections contained in a notebook, and the pages that are in a section. To view an item, all you have to do is tap it and it will be displayed below all pinned items. The Recents tab creates a pin icon on the right of every item you have viewed. By tapping this icon, you can move the item to the top of the list regardless of the item that you have touched. The item that you pin next will go below the top item and so on.

- The Back Button: The back button appears in the upper left corner of the display on every window except the home page. The button resembles a left facing arrow, and tapping it will take you to any section, note or list that you were viewing previously. Depending on where the button will take you and where you are in the notebook, the text that accompanies the back button will change. For instance, if you are viewing notebook sections, the button will display home, however, if you are displaying a particular page in a section of a notebook, it will display the name of the notebook. It is important to remember that regardless of the text that the button displays, its primary function is a back button.

- New Section Button: This button appears next to the back button everywhere except on the home screen

- View Icons/Search Button: The bottom of the window contains three buttons that you can tap to change what you view on the display. The first changes the view of the list pane and shows you any unfiled notes that you may have, the next button displays recent notes, and the third brings up the search view.

- Keyboard: once you have opened a note, when you click on the pane a keyboard will appear automatically. The keyboard on iOS devices has its own functions including a check box button that you can tap to add check boxes to items that you have selected, and a bullet button that adds bullets to the selected text. It also contains a camera button that allows you to take photos to add to a note, or to add pictures to the note, and a return button that changes depending on where you are in the app. For instance, if you are trying to log in, the button will display Sign In, while if you are trying to search for something, the button will read Search rather than Return. If you would like to close the keyboard, just tap on the close keyboard button on the bottom right on of the keyboard.

Creating and Editing Notes in One Note for iOS

When it comes to creating, writing and editing notes on One Note for iOS, the process has been made as simple as possible, though you must remember that the app has fewer features than its online and desktop counterparts.

Creating a new note is simple and can be done regardless of where you are in the application. All you have to do is tap the new note icon in the upper right hand corner of the window to reveal a pop up menu with two options, Create Note (Unfiled) or Create Note in Current Section. The first choice creates a new note in your Web or Personal notebook and stores it under Unfiled Notes, while the second option creates a note in whatever section displayed at the top of the window, regardless of where you are in the application. If you are actually viewing the Web notebook's unfiled section and you click on the New Note option then the second option will be unavailable because it is basically redundant.

If you would like to open an existing note from Microsoft One Drive that is relatively simple as well. All you have to do is access the home screen and tap the name of the notebook, the name of the section, and the name of the note that you wish to access.

Adding a picture is also very simple, and can be done by tapping on the camera icon at the top of the keyboard. You will then be presented with two options:

1. Camera: Tapping on camera will activate your iOS device's camera, allowing you to take a picture that you can then add to your note. The camera interface is relatively easy to understand, and even has a button that can change the camera you are using between the rear camera and the front camera. After you take your picture, two icons will appear at the bottom of the screen, the one of the left allows you to discard the photo you just took and take a new one, while the one on the right confirms that the photo you have been taken is the one that you would like to insert into the app.

2. Photo Library: Tapping on this option will allow you to choose an item from your device's photo library

If your device does not have a camera, then you will be presented with two different options, Saved Pictures and Photo Stream.

When it comes to renaming notes and notebooks, the procedure is much the same as with the One Note for Android App, with many of the same restrictions. This means that you can edit the name of a note, but you cannot edit the name of a section or the whole notebook without logging on to the One Note web app or desktop application. You also cannot create a new notebook in One Note for iOS.

However, you can delete notes in the iOS app using one of two procedures

1. Swipe the note item to the left in the list pane. A red delete button will appear, and if you are sure that you would like to delete the note then you can tap delete and the note will disappear.

2. Tapping the trash can icon at the top of the display will bring up a menu with an option to delete a page, if you are sure you want to delete the

page tap the Delete This Page option and the note will disappear.

Searching through Notes in One Note for iOS

Searching through notes in One Note for iOS is much the same as searching through notes on the web and desktop applications, and the search feature allows you to search for single terms in the notes contained in all the different notebooks. However, due to the limited functionality of the app compared to the desktop version, you can only search for text within notes, and you cannot search for pictures or tags.

Searching for text in One Note for iOS is as simple as tapping the search button at the bottom of the screen and type the search term into the Search Notebooks field at the top of the list pane. The results of your search should appear in the list pane below the search field. Once you have found the note that you would like to edit, you can view it in the note pane by tapping it. You will not lose any of your search results when you do this as the list pane will not be affected by your tapping on the note.

Managing notebooks on One Note for iOS app has been made extremely difficult, and the only way to effectively rename, delete or move notebooks is through the One Drive app for iOS.

Configuring One Note for iOS

One Note for iOS has very few settings that you can actually modify. To change the settings that are available to you in the app, the first thing you must do is tap the settings button at the top of the home screen. The window that opens has a number of options including:

1. Upgrade: Like One Note for Android, One Note for iOS limits the number of notes that you can have to 500. However, you can have unlimited notes if you pay a nominal fee to upgrade the app.
2. Sync Now: Tapping on this item immediately syncs all notes that are programmed to sync automatically
3. Notebook Settings: This button allows you to choose which notebooks to sync automatically while simultaneously allowing you to choose which notebooks will be displayed on the home screen
4. Image Size: A feature that is notably absent from its Android counterpart, the image size setting allows you to select the size of the images that you insert into different notes
5. Sign Out: Tapping this button will sign you out of One Note
6. Help and Support: Tapping this button will open a window that gives you links to different community support forums and help documentation
7. Terms of Service: Tap this button if you would like to view One Note's terms of service
8. Privacy: Tap this button if you would like to view One Note's privacy settings

Just like with its Android counterpart, One Note for iOS has some settings that you cannot change within the app, but that you can change using your device's settings. To access these settings, tap on the settings icon on your device, and tap on the One Note icon in the settings pane. The settings that appear are outlined below:

1. Sync on Wi-Fi Only: This particular option ensures that notebooks will only sync with One Drive when you are within Wi-Fi range. This allows you to keep the data usage of the app to a minimum, and ensures that you do not go over your data limit.
2. Reset Application: This feature is especially useful when you have sync errors, or other errors that will not fix themselves regardless of how many times you restart the application. Once activated, you will have to sign yourself back in to One Note, but you will not lose any information
3. Version: There is nothing you can do to change this setting as it is just a description of the version of the app that you currently have installed on your device.

Managing Syncing and Images in One Note for iOS

Setting individual notes to sync automatically is something that may seem unnecessary, but is one of the most useful features of One Note for iOS, especially if they are set up so that they sync regardless of whether you can see them on the home screen or not. The steps below outline how to turn OFF automatic syncing for particular notebooks, though following the opposite of this procedure will turn ON syncing for any notes that you may have created:

1. When you are viewing the home screen, tap the settings button at the top of the screen and tap on notebook settings. A pop up window should appear
2. Tap on the on slider that is next to a notebook to turn syncing off for that particular note. Once the slider is in the off position, you will not be able to view the notebook on the homepage, and the note will not sync automatically.
3. Close the settings pane by tapping on the settings button and then tapping the close button in the upper right hand corner of the window. Once you return to the Home screen, you will realize that the notes that you have toggled off are no longer visible, and they will not sync automatically.

Managing image settings on One Note for iOS is simple despite the fact that there are quite a few options to choose from when it comes to how you would like the app to display photos. For instance, you can choose from a wide range of image sizes when you are inserting the images to notes:

- Small sets the images to .5 megapixels
- Medium sets the images to 1 megapixel
- Large sets the images to 2 megapixels
- Actual Size will display the photo in notes at their actual sizes

- Ask me will ensure that you receive a prompt every time you try to insert an image in a note so that you can determine for yourself how large the image should be when it is being inserted into the note.

Chapter 10: Top 10 Tips for One Note Users

There are many tips and tricks that have been outlined in this book that will make using One Note that much simpler. There are also many scenarios that have been described that show you how useful the application can be. This chapter aims to emphasize some of these points, while simultaneously give you a few tips which could make your use of the program easier.

1. Backing up Important Data

Syncing and backing up of notes and notebooks has been discussed at length in various sections of this book, but the importance of this feature needs to be emphasized one more time so that you can truly understand how useful the feature can be.

Consider this scenario, you are on your way to Europe from the West coast for a well-deserved vacation and need to catch a connecting flight in New York. However, when you arrive in New York, you realize that you have lost your hand luggage, and that it had all the information you need for your trip. However, you still have your cell phone, and you remember that your itinerary was saved in One Note, which you can access through your phone. Because of the data backup, you make it to your next plane with barely a moment to spare, but you find it, and you're on your way to Barcelona. Another scenario where you could be saved by the backup feature on one note this, you are on your way for a job interview, but you forget to bring and review your resume to the meeting. However, you have backed everything up on One Note and you have your iPad with you. You go for the interview anyway, and when you are asked for your resume, you hand over your iPad with all the information that your future employer needs on it. You never know, you may ace the interview purely because you have your resume on an iPad.

Perhaps you were on your way for a meeting at work and you forgot to carry that report that was due today because you were more occupied with chowing down breakfast and getting out of the house as fast as you can. Now you are two hours away from home and the meeting is in an hour. Then you remember that you have the presentation on One Note and all you have to do is print out the hard copies and connect your phone to the projector and you're sorted.

The number of scenarios that One Note could help with are limitless, but the examples outlined above should be enough to show just how important the application can be.

2. You Have Access To Entire Office Documents On Your Phone At

The Touch Of A Button
One Note is one of the only office apps that allows you to access documents easily on Android. Despite the fact that you may not be able to work much on your documents, you will at least be able to access and view them on your mobile device.

This feature is especially useful when you realize that saving the document as a printout on One Note rather than pasting it in to the application will allow you to view it exactly the same way you would in the original program. You may not be able to edit the documents, but you will at least be able to view them perfectly in One Note. If you have the One Drive application on your phone, you can access Word, Excel and PowerPoint documents via the appropriate web applications and actually edit the files, though you will have limited features compared to the full office applications.

3.   Dictating Notes to Text

This is another feature that has been addressed in earlier chapters, but its importance cannot be emphasized. Though this feature is most useful in One Note for Android as it does not allow you to store recorded voice notes, it is a very useful option that should be seen as an added feature rather than a missing option, especially if that was your aim in the first place. This feature is especially useful for those people that have to translate a lot of dictated audio into plain text.

4.   Retrieving Text from Images

This feature has been discussed at length in previous chapters, but its importance cannot be overstated, especially for those people who work with images on a regular basis. The fact that it is so easy to do is one of the best things about this feature, however, it is important to note that you have to ensure that your photograph is of the highest quality possible to ensure that the text that you are trying to grab from the image comes out as clearly as possible.

You must also ensure that the photo is not too dark or too bright, and that the font in the picture is not one of those difficult to read calligraphic fonts that are sometimes used, especially by creative individuals. However, in most cases, even if the picture is dark or the font is hard to read, you will still be able to retrieve at least some of the text, meaning you will be able to avoid most of the retyping that you need to do.

5.   Taking a Screengrab and Marking it

As was mentioned previously, one of the best things about One Note is that you can take a screen shot of something using the clipping tool in the application. You can then use the pen tool in One Note to mark up the image. This is a very useful feature especially if you would like to add notes and comments to a photograph or image and share it for the world to see. For instance, if you are in the middle of developing a website and you would like to show your contributors what you would like to keep or change in the website, you can use the pen tool to note anything that you feel should be noted.

If you do not have a computer to a PC that has digital pen technology, you could always use actual text, or you could even use your mouse pointer as a pen and draw your comments if your handwriting with the mouse is good enough.

6. Marking Up Documents with a Pen

Should you be able to access a digital pen-capable PC, you will realize that marking up documents in One Note is easy. However, if you would like to ensure that you keep some of the similarities to the original document's formatting, then you will find it is easier to import a printout of the document instead of copy/pasting the contents of a page or document onto the note page.

If you have access to other Microsoft Office applications on your pen-capable PC then there are some cases where you could use the pen to mark up the documents in that app in much the same way. However, you must keep in mind that it can be trickier to do so in those apps, so it may be easier to just stick to One Note when marking up documents.

7. Copying links into Paragraphs

If you ever need to create links to specific parts of a One Note page, you can do it easily by right-clicking on the paragraph or note container that you would like to link and clicking on the Copy Link to Paragraph option. This will allow you to copy a link to your computer's clipboard, and you can then add that link to any document of your choice.

8. Searching Text In Images

This is another feature that has been addressed before but cannot be emphasized enough. The ability to search for text in images means that any image that you drop into One Note that has text in it can be discovered as long as you make the text in the image searchable. Even better, you can specify the language that the text is in so that you can find it later. This will allow you to find images quickly regardless of the language that the text in the photograph is in.

9. Docking One Note to the Desktop

If you find yourself using One Note regularly, you can make sure that it is always within easy reach by placing it beneath all your other open windows. This will ensure that you do not always have to select it from the taskbar, and that you can edit your notebooks whenever you need to without much hassle.

To dock One Note to your desktop, all you have to do is click on the view tab and click on the Dock to Desktop option and the app will automatically dock itself to the desktop below all the other open windows. Other windows may move a little, but you will always be able to see all the information displayed in them, as well as the information displayed in One Note. However, it must be noted that this feature works best if you have a large monitor screen as resizing the windows may make text hard to read.

10. Creating Outlook Tasks in One Note

One of the most useful features of One Note is the ability to create Outlook tasks that can then be modified at a later date. These tasks are created and edited via a drop-down menu in One Note. The menu has a range of options that include options to delete or open a task in Microsoft Outlook. The tasks that you create will add themselves to your Outlook automatically as long as you have set up and configured the application on the same computer that you are using One Note on.

It is important to remember that you do not have to create a new task then add the task information to it. It is much simpler to choose an existing item and then select an option from the drop-down menu that is displayed on the screen. Once you have selected the appropriate option, the item will immediately be made a task.

CONCLUSION

OneNote is one of the most versatile note taking applications on the planet, and is well suited for a wide variety of people from doctors and lawyers to students and researchers, to those people who just love to write things everywhere so that they will not forget them.

If you are one of those people who has never experienced a program or application like this then be warned, this program is addictive. The countless features and endless possibilities will keep you staring at the screen for hours, regardless of whether you are using your smartphone, laptop or PC.

The way it integrates seamlessly with the other programs in the Microsoft Office Suite is also very impressive and is something that you will soon be taking for granted once you get comfortable using the program.

Though OneNote is defined as a free-form information gathering computer program, it is way more than that. It is an organizer, a spreadsheet application, a database management system, a word processor, and countless other things rolled into one.

Although it is not geared towards publishing work, because it integrates so well with the rest of Microsoft Office you do not really need to worry about that. What you do need to pay attention to is the fact that it will make doing your work much easier, and when you feel you have done enough to publish it, you can just transfer it to one of the other programs in the Office Suite to polish it off before releasing it to the world.

This manual has given you a peek into the world that is OneNote. It is a world that is ever changing and full of possibility, and if you choose to enter it, you will never be disappointed.

www.ingramcontent.com/pod-product-compliance
Lightning Source LLC
Chambersburg PA
CBHW030442220526
45464CB00006B/2382